PRACTICAL POWER ELECTRONICS

PRACTICAL POWER ELECTRONICS

APPLICATIONS, EXPERIMENTS AND ANIMATIONS

MUSTAFA HUSAIN

PARTRIDGE

A Penguin Random House Company

To order additional copies of this book, contact
Toll Free 800 101 2657 (Singapore)
Toll Free 1 800 81 7340 (Malaysia)
orders.singapore@partridgepublishing.com

www.partridgepublishing.com/singapore

In memory of late Elbaz Ayad

Work. Finish. Publish.
Michael Faraday (1791-1867)

PREFACE

Why another book on power electronics? It is not just another book. Most of the published books in this area focus on the design aspects and they are full of mathematics that distracts the learners from understanding the operating principles which practicing engineers and technicians need in their daily jobs. Moreover, when it comes to experiments, the available books are scarce.

We know of many electrical engineering graduates who passed a power electronics course but, when they joined the industry, they saw a big gap between what was taught and what knowledge and skills the industry needs. For instance in a power electronics course, all graduates should know the symbol, operation and applications of the thyristor. But many of them might not touch or see the physical appearance of the thyristor nor might they see industrial applications of the thyristor. This would be quite a shock if these graduates caught jobs that happen to involve the operation and maintenance of thyristors in big systems. How can we prepare learners for such jobs?

This book aims to be the first step in narrowing the gap between academia and industry. It is an organized collection of useful notes that we have used for 15 years to teach the basics of power electronics theoretically and practically. We let our students touch and feel power electronics. We do practical demonstrations and we dismantle appliances and panels to show them the physical appearances of devices and controllers that work together to form a power electronics converter. We have different shapes and sizes of devices and converters manufactured by a variety of global designers. In addition, we organize industrial visits to see the applications and operation of power electronics. Instructors are strongly advised to carry with them power electronic devices and show them to their classes and also to give their students the opportunity to touch and feel the devices.

It is wise to keep these notes in a book and it is a pleasure to share them with interested instructors and professionals worldwide. Being concise, it aims to explain the basic operating principles briefly and to present some applications of power electronics without deep explanation.

The book is divided into two parts; theoretical and experimental. The first part consists of eight chapters covering the operation of devices and circuits. The basic converters are ac-dc, ac-ac, dc-ac and dc-dc converters. Some applications of these converters are listed and some are briefly explained. The second part has six experiments that will help the learner have the feeling of power electronics and to understand the concepts practically. Included also is the employment of microcontrollers to provide simple control of converters.

To assist the learner further, photos of power electronic devices and circuits of some gadgets and appliances that we use daily are included. The text is enriched with simple diagrams. For easier and quicker understanding, some concepts were animated. The animations are available on EduMation.Org, a website dedicated for educational animations. The animations may also be found on the EduMation channel on YouTube. Data sheets of power electronic devices are attached at the end of the book.

This book is valuable for teaching a first hands-on course on power electronics for technical institutes, polytechnics and in-house training for industries. Instructors will find it easy and straightforward while preparing for their classes.

I shall express my gratitude to the kindness and support offered by the following persons. My colleagues at Bahrain Training Institute: The late Mr. Elbaz Ayad always corrected mistakes and provided invaluable comments; his departure was a great loss to our team. I am indebted to Mr. Abdulhusain Buhusain for his team spirit and restless encouragement. I feel very grateful for Mr. Hassan Alkhayer who prepared the template of the manuscript. Special appreciation goes to Mr. Fayez Abbas who constructed and tested a microcontroller-based on-off ac-ac converter. Many thanks to our experienced colleague Mr. Yosri Idrees who dismantled power thyristors from an old UPS system, and he kindly allowed us to use them during power electronics courses.

I am indebted to Powerex Inc who gave me permission to use photos and datasheets of some of their products. My appreciation is also extended to ON Semiconductor Inc and Diodes Inc for their datasheets.

The manuscript was proofread by Ethra for Translation and Research; they thankfully provided valuable comments as well. The book would not have been started without the support of my wife and children Ali, Kawthar and Mahmood.

Sharing comments and ideas with you will help me improve the book. If you have comments or suggestions, please do not hesitate to send them via email or WhatsApp. I will be glad to hear from you.

Dr. Mustafa Husain
Founder- EduMation.org
Mobile/WhatsApp: +973 3557 9902
Email: msdaif@gmail.com

CONTENTS

PART I
THEORY

PART II
EXPERIMENTS

PART I

THEORY

1

INTRODUCTION

1.0 Objectives

Power electronics are devices made of semiconductor materials the most common of which are silicon and germanium. Their main function is switching, and thus they are utilized for electrical energy conversion and control. Many converters can be built with power electronics such as rectifiers, inverters, choppers and so on. They are of great importance to the electric power industry worldwide.

Power electronic devices differ from electronic devices in their power ratings. Hence, any electronic device can be classified as either a signal or power device. A signal electronic device is used for low-power applications requiring low voltage and current. A power electronic device withstands high voltage and current.

We are concerned with power electronic devices and systems but the knowledge of some signal devices may be essential. We will study the basic principles, circuits and applications.

1.1 History of Power Electronics

After WWII, a research was started at Bell Labs in the USA to replace thermionic valves that were bulky, inefficient and of a short life. Semiconductors such as silicon and germanium were proposed. After smart thinking and hard work, semiconductor electronics saw light in December 1947. John Bardeen, Walter Brattain and William Shockley invented the contact-point transistor as an amplifier. In 1956, they received the Nobel

Prize in Physics for this breakthrough. A replica of the first transistor manufactured by Bell Labs appears in Fig.1.1. It has two gold contacts pressed on a high-purity germanium slab.

The transistor revolutionized the electronics industry, and it gave birth to the information technology that we use nowadays. Another milestone is the invention of the thyristor in 1958 which urged General Electric to produce commercial products known as SCRs. The TRIAC and GTO were introduced in the 1960's whereas the power MOSFETS and BJTs came out in the 1970's. Many other devices were invented and the process is still going on.

1.2 Classification of Power Electronics

The power electronic devices may be classified in terms of their turn on/off characteristics as:

- Uncontrolled turn on/off: an example is the diode because its turn on/off cannot be controlled.
- Controlled turn on/off: the BJT, GTO, MOSFET and IGBT can be turned on and off simply.
- Controlled turn on and uncontrolled turn off: the SCR can be turned on via its gate but it cannot be turned off in the same manner; some schemes may be implemented for the turn-off process.

The gate requirements of these devices may be classified as:

- Continuous gate signal: BJT, MOSFET and IGBT
- Pulse gate signal: SCR and GTO

Some devices withstand reverse voltage or current polarity and they can be classified as:

- Unipolar voltage capability: BJT, MOSFET, GTO, and IGBT
- Bipolar voltage capability: SCR and GTO
- Unidirectional current capability: Diode, SCR, GTO, BJT, MOSFET and IGBT
- Bidirectional current capability: TRIAC

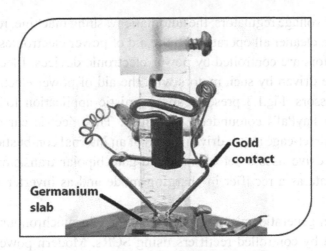

Fig.1.1 Replica of first transistor. © Mark Richards

1.3 Converters

We will study different types of circuits which are called converters. A converter is a circuit that converts electric power from one form (AC or DC) into another (AC or DC). Any course in power electronics covers the following converters:

- Uncontrolled Rectifiers: using diodes, power can be converted from fixed AC into fixed DC.
- Controlled Rectifiers: using SCRs, power can be converted from fixed AC into variable DC.
- Inverters: this is a DC-AC converter using SCRs, BJTs or MOSFETS.
- Choppers: this is a DC-DC converter where DC voltage can be stepped up or down.
- AC Voltage Controller: this is an AC-AC converter. Using SCRs, the power flow can be controlled by controlling the phase angle of the voltage signal.

1.4 Applications

Power electronics have changed the world and there are countless applications. A microwave oven would contain a high-voltage diode, a

TRIAC and voltage regulators; the automatic washing machine, refrigerator and vacuum cleaner all operate with the aid of power electronics.

AC motors are controlled by power electronic devices. Electric trains (Fig.1.2) are driven by such motors with the aid of power electronics and microprocessors. Fig.1.3 presents an optimistic application foreseen and invested by PayPal's cofounder Elon Musk. This electric car employs a 3-phase squirrel-cage motor drive instead of an internal combustion engine. The motor drive uses a set of insulated-gate bipolar transistors (IGBT) which operate as a rectifier in charging mode and as inverter in driving mode.

In power generation plants, the field system of a synchronous generator is regulated by controlled rectifiers using SCRs. Modern power supplies employ many devices. Telephone companies require their power supply available for its computer systems and telephone exchanges. Power electronics are utilized to make systems that maintain the flow of electric power to these systems with the aid of Uninterruptible Power Supply (UPS). Nowadays, UPS systems are used in hospitals, money exchanges, airports, etc.

Fig.1.2 Power electronics are employed to drive this train. © Dubai Metro

Fig.1.3 Tesla Motors' Model S being charged, a modern electric car. The liquid-cooled power train consists of a 60 kWh Li-ion battery, squirrel-cage motor with copper rotor, drive inverter and a gearbox. The car can be controlled from a touch screen. © Tesla Motors Inc

2

DEVICES

2.0 Introduction

The basic power electronic devices are covered in this chapter. All of these devices are static switches but they differ in how they are turned on or off. They are also different in their maximum power ratings and their speed of operation. These devices are manufactured in discrete or module forms. There are different types of packages for each device.

2.1 PN-Junction Diode

The diode is the basic semiconductor device usually made of silicon, and there are several types of diodes. The pn-junction diode is the mostly used device in power applications especially in rectifiers. It is also used in home appliances, in automotives, in motor drives, in welding machines, etc.

Construction: Several power diodes are shown in Fig.2.1. They are manufactured in different packages some of which are: DO-41, capsule and stud-mount type. The diode consists of two terminals: anode (A) and cathode (K). The symbol of the diode is shown in Fig.2.2. The anode is the positive terminal and the cathode is the negative terminal. The polarity here does not necessarily imply that the anode must be connected to a positive polarity point or the cathode to a negative point. The polarity is used for indicating the turn-on/off process.

Operating principle: When the anode is more positive than the cathode, the diode turns on and acts like a closed switch (Fig.2.3a). The diode turns off when the cathode is more positive than the anode (Fig.2.3b). The diode breaks down when the cathode is at a very higher positive potential than the anode.

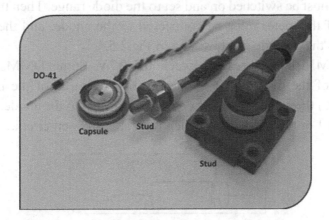

Fig.2.1 A variety of diode packages. © Mustafa Husain

Fig.2.2 Symbol of pn-junction diode.

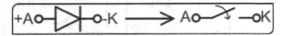

a) Diode turns on when anode is more positive than cathode.

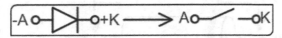

b) Diode turns off when cathode is more positive than anode.

Fig.2.3 Diode switching conditions.

Testing with Digital Multimeter (DMM): Testing a power diode with a DMM is very simple. It helps us identify its terminals and guide us in knowing its working condition. Testing with the DMM is not always conclusive. More accurate judgment can be made by building a circuit of a dc source and a lamp connected in series with the diode in forward and reverse connections.

Let us assume we want to test a diode of the 1N54xx series the physical appearance of which is drawn in Fig.2.4. To test the diode with a DMM, the DMM must be switched on and set to the diode range. Then the positive terminal of the meter must be connected to the anode, and the common terminal to the cathode as illustrated in Fig.2.5.

The DMM should read between 0.5-0.7V. Some DMMs read the resistance of the diode in Ω. If the positive terminal of the meter was connected to the cathode and the common terminal to the anode, the meter would read 1 or OL (over load) which indicates an open circuit.

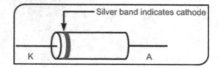

Fig.2.4 Physical appearance of 1N5401 diode.

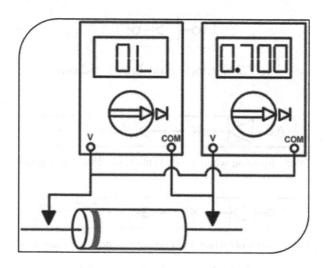

Fig.2.5 Testing the diode with a DMM.

Example 1: Basic understanding

Are the following silicon diodes on or off? Justify your answers.

a)

$$+3V \hspace{0.5em} \rightarrowtriangle\!\!\vdash \hspace{0.5em} +1V$$

b)

$$+5.5V \hspace{0.5em} \rightarrowtriangle\!\!\vdash \hspace{0.5em} +5V$$

c)

$$-3V \hspace{0.5em} \rightarrowtriangle\!\!\vdash \hspace{0.5em} +1V$$

Solution

a) On because the anode potential is more positive than the cathode potential by at least 0.7V.
b) Off because the anode potential is not more positive than the cathode potential by at least 0.7V.
c) Off because the anode potential is negative.

Example 2: Advanced understanding

Describe the turn-on process of the diodes D_1 and D_2 shown below at all intervals 0-t_2.The anode potential of D_1 is v_1 and that of D_2 is v_2. The cathodes are connected to a resistor R_L.

Solution

From time $t=0$ to $t=t_1$, the anode potential of D_1 is negative and the cathode potential is zero. Hence, D_1 is off. At the same time, the anode potential of D_2 is positive and so it is on.

From time $t=t_1$ to $t=t_2$, D_1 switches on because the potential of its anode rises to positive values. During this interval, D_2 switches off as its anode goes through a negative potential.

2.2 Thyristors

One drawback of the diode is its uncontrolled operation. It automatically turns on or off depending on the potential values of the anode and cathode. It is advantageous to make a device which can be operated at will; the thyristor was invented for this purpose.

There are different types of thyristors depending on the method of switching: SCR, GTO, IGCT and TRIAC. Like the diode, the thyristors are widely used especially in motor drives, electric trains, HVDC, electric heating control, electric welding, UPS and so on. There are two main types of applications depending on the speed of switching:

- Phase-control thyristors
  Used in low-speed switching applications. They operate at the line frequency.
- Inverter thyristors
  They are fast-switching devices which are used in choppers and inverters.

2.2.0 Silicon Controlled Rectifier (SCR)

The SCR is the first thyristor type to be invented. Until today, it is the only device that can handle the highest power.

Construction: Two industrial SCRs are shown in Fig.2.6. The data sheets are attached in the appendix. The SCR consists of three terminals: anode (A), cathode (K) and gate (G). The SCR symbol is shown in Fig.2.7.

Capsule package

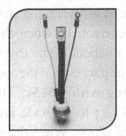
Stud-mount package

Fig.2.6 Industrial SCRs. © Powerex Inc

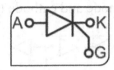

Fig.2.7 Symbol of SCR.

Operating principle: The SCR can be turned on when the anode potential is more positive than the cathode potential, and when a positive current pulse of short duration is injected into the gate as illustrated in Fig.2.8. Once turned on, it cannot be turned off through its gate. It turns off when the cathode potential becomes more positive than the anode potential (Fig.2.8b).

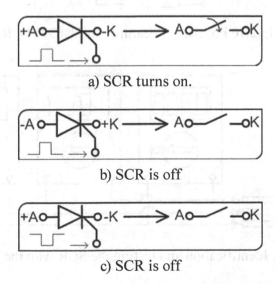

a) SCR turns on.

b) SCR is off

c) SCR is off

Fig.2.8 SCR switching conditions

It also turns off when the current (from anode to cathode) drops below the holding current, or when the direction of the current becomes negative (cathode to anode). The thyristor does not turn on by applying a negative pulse to the gate even if the anode is more positive than the cathode (Fig.2.8c).

Testing with DMM: The equivalent circuit of the SCR is shown in Fig.2.9. The gate-cathode junction can be tested with the DMM just like testing the diode. When the positive terminal of the DMM is connected to the gate and the common terminal to the cathode, the meter will read around 0.7V if the SCR is OK. It will read open if the SCR is damaged i.e the gate-cathode junction is damaged. The meter will also read open if its terminals are connected between the anode and cathode (Fig.2.11).

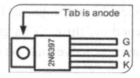

Fig.2.9 Diode equivalent circuit of SCR

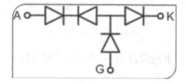

Fig.2.10 Physical appearance of 2N6397 SCR

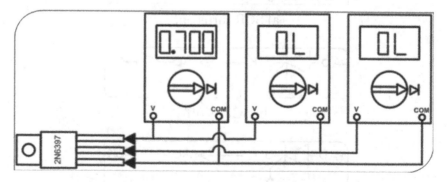

Fig.2.11 Identification and testing the SCR with the DMM

2.2.1 Gate Turn Off (GTO)

The GTO thyristor is a modification to the SCR. It is favored over the SCR for many applications.

Construction: It also consists of anode, cathode and gate. The symbol of the GTO is slightly different (Fig.2.12). The two arrows show the bidirectional operation of the gate.

Operating principle: Like the SCR, the GTO can be switched on by applying a positive current to the gate. Unlike the SCR, this gate current should be continuous to ensure reliable operation of the GTO. The GTO can be turned off by applying a negative current pulse to the gate. The drawback is that the negative gate current is quite big compared to the triggering current.

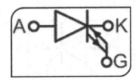

Fig.2.12 Symbol of GTO.

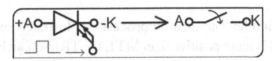

a) GTO turns on by a positive pulse.

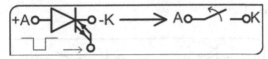

b) GTO turns off by a negative pulse.

Fig.2.13 Operation of GTO.

2.2.2 Triode AC (TRIAC)

The SCR and GTO are unidirectional switches and so they would pass only a half cycle of a sine wave; in this case, half of the power would be utilized. Bidirectional switching is necessary to control the full ac waveform.

The TRIAC is a bidirectional switch that is used in heating and lighting applications. Light and fan controllers employ the TRIAC to control the intensity of light and the speed of the fan.

Construction: The TRIAC is a three-terminal device consisting of two anti-parallel SCRs but with a single gate. It is packaged in a single chip. Its terminals are: main terminal 1 (MT1), main terminal 2 (MT2) and gate (G). The symbol is two anti-parallel diodes with a gate.

Fig.2.14 2N6073 TRIAC.

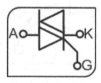

Fig.2.15 Symbol of TRIAC.

Operating principle: The turn-on process is illustrated schematically in Fig.2.16. If MT2 is more positive than MT1, the TRIAC can be turned on by applying a positive gate pulse with respect to MT1. If MT1 is more positive than MT2, the TRIAC can be turned on by applying a negative gate pulse with respect to MT1.

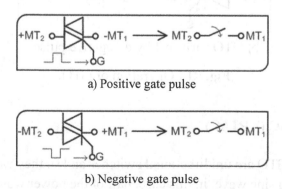

a) Positive gate pulse

b) Negative gate pulse

Fig.2.16 TRIAC switching conditions.

2.2.3 Light-Activated Devices

A photo SCR is triggered by applying light onto its gate. Hence, it has only two terminals. The light is often provided by an LED, and the SCR and LED are packaged into one device known as solid state relay (SSR). It is used in high power applications where the firing circuit is of low power and isolated from the power circuit. The power circuit is of high power and the rating of the SCR could be in kW but the rating of the triggering circuit could be in mW.

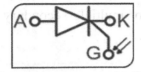

Fig.2.17 Symbol of photo SCR

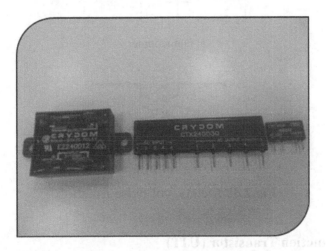

Fig.2.18 An assortment of SSR. © Mustafa Husain

2.2.4 Thyristor Firing and Protection Circuits

In thyristor circuits, the gate circuit is the control circuit whereas the power circuit is between the anode and cathode. The power circuit is at higher voltage than the control circuit. Typical values for power circuits are higher than 100V, and 12 to 30V for control circuits.

The control circuit must be isolated from the power circuit. This can be achieved by the following methods:

- Opto-coupler: An opto-coupler or opto-isolator employs an infra-red light emitting diode (ILED) which forms the input and a photo semiconductor device which forms the output. A pulse applied to the input of the coupler turns on the diode whose light activates the gate of the photo device. The photo device could be a BJT, Darlington, SCR or TRIAC.
- Pulse transformer: This is a 1:1 transformer with one primary winding and one or multiple secondary windings to be fed to a number of devices.

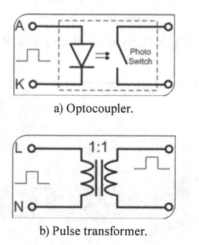

a) Optocoupler.

b) Pulse transformer.

Fig.2.19 Methods of pulse isolation.

2.2.5 Unijunction Transistor (UJT)

The UJT is commonly used to generate triggering pulses for the SCR and TRIAC but it may be used to trigger other devices. It has three terminals: emitter E, base one B_1 and base two B_2.

The UJT switches on when an adequate voltage V_E appears between E and B_1. A pulse generator or relaxation oscillator employing a UJT is shown in Fig.2.20. The capacitor charges through R until its voltage reaches a peak value that is enough to turn on the UJT. This is so because the capacitor voltage is the same as the E- B_1 voltage.

When the UJT is on, the capacitor discharges through R_{B1} until its voltage is no longer enough to bias the transistor and so the UJT switches off. This process repeats and the voltage across R_{B1} represents the charging and discharging process of the capacitor. The triggering pulses across R_{B1} can be fed into the gate of a thyristor.

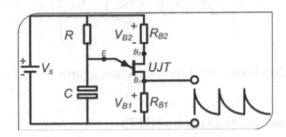

Fig.2.20 Pulse generation with UJT circuit.

2.2.6 Diode AC (DIAC)

The DIAC is two anti-parallel connected diodes but assembled in one chip. It is often used to provide triggering pulses for the TRIAC especially in light dimmer circuits. Unlike the diode, It can conduct in both directions and its break-over voltage, V_{BO}, is much higher than the forward voltage of the diode. A typical value of V_{BO} is 30V. For higher values like 250V, the DIAC is named differently: SIDAC.

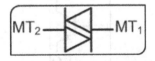

Fig.2.21 Symbol of DIAC.

2.2.7 Protection Against di/dt and dv/dt

The SCR may be damaged if its current rises too quickly (*di/dt*) or if the voltage across it builds up to high values in a short period of time (*dv/dt*). The SCR has to be protected against high di/dt rates by incorporating an inductor in series with the device, and an RC circuit is connected across the device for dv/dt protection (Fig.2.22). The RC circuit here is called a snubber.

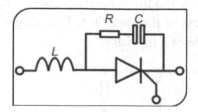

Fig.2.22 Snubber circuit for protection against di/dt and dv/dt.

2.3 Bipolar Junction Transistors (BJT)

The BJT transistor switches on and off faster than the SCR and thus they are used in high frequency applications as high as 50 kHz but their power capability is lower than the SCR. The BJT is a current-controlled device that comes in two types: NPN and PNP. The NPN transistor is mostly used in power electronic applications and so it will be studied here.

2.3.0 NPN Transistor

It has three terminals: base (B), collector (C) and emitter (E).

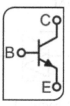

Fig.2.23 Symbol of NPN transistor.

Operating principle: The transistor can be turned off when the base current I_B is zero or when it is not sufficient. It turns on when I_B is sufficiently large depending on the value of the collector current I_C.

The base should be at a higher positive potential than the emitter. When the transistor is on, there is a voltage drop across it defined as V_{CE}. This voltage drop is practically 1-2V for power transistors. The flow of the collector current and the occurrence of the voltage drop results in losses in the form of heat.

- V_{CEO}: The maximum allowable collector-to-emitter voltage when the base terminal is left open circuit.
- h_{FE} (min) @ I_C: It is the dc current gain and it is defined as the ratio of dc collector current to dc base current at the stated collector current or:

$$h_{FE} = \frac{I_C}{I_B} \quad \textbf{Eq.2.1}$$

The value of h_{FE} increases as temperature increases, and it increases as collector current increases. The transistor turns on when

$$I_C > h_{FE} I_B \quad \textbf{Eq.2.2}$$

2.3.1 Darlington Transistor

It is commonly applied to drive relays and solenoids where it receives pulses from a PLC, a microprocessor or logic circuits. In such applications, it is favored over the discrete transistor because it needs lower base current to turn on and it can handle higher power.

Construction: A Darlington transistor is made up of two BJT transistors connected as depicted in Fig.2.24. The pair comes fabricated in one package but a Darlington can be formed by two discrete transistors. The purpose of the Darlington configuration is to have higher gain and higher power.

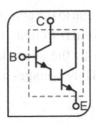

Fig.2.24 Symbol of Darlington transistor.

2.4 MOSFET

The *Metal Oxide Semiconductor Field Effect Transistor (MOSFET)* is a voltage-controlled transistor. It is a solution to the breakdown phenomenon that is often experienced with the BJT. It is very fast and hence it is incorporated in applications requiring high frequency (up to 1 MHz) and low power (up to a few kW's) such as inverters and choppers.

Construction: There are two types according to the direction of currents: N-channel and P-channel. It consists of a gate (G), drain (D) and source (S) and the symbol of an N-channel MOSFET is shown in Fig.2.25.

Fig.2.25 Symbol of MOSFET.

Operating principle: The transistor can be biased by a positive voltage between the gate and source (V_{GS}). Current flows from the drain to the source and so a voltage drop V_{DS} develops across the device. The transistor is now switched on, and it can be switched off by removing the gate signal.

2.5 Insulated Gate Bipolar Transistor (IGBT)

The BJT can easily breakdown due to the connected junctions in its structure. The IGBT consists of a gate similar to that of the MOSFET. The insulated gate makes it less vulnerable to the secondary breakdown phenomenon common to the BJT. The IGBT is voltage-controlled. It is faster than the

BJT but slower than the MOSFET but it has higher current densities than power MOSFETs. Therefore, they are more cost-effective in many high power, moderate frequency applications.

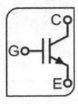

Fig.2.26 Symbol of IGBT.

3

DIODE RECTIFIERS

3.0 Introduction

A rectifier converts fixed AC voltage into fixed DC voltage. This is known as uncontrolled rectification because the output DC voltage cannot be varied. There are several configurations of diode rectifiers. Most rectifier circuits are used as DC power supplies and thus they employ step-down transformers. They are also applied for the operation of DC motors.

3.1 Half-Wave Rectifier

This rectifier consists of a single diode and it is the simplest type of rectifiers. A common application is the high voltage circuit for a microwave oven which appears in Fig.3.1. It consists of a single diode, a capacitor, a transformer and a magnetron. The transformer steps up the mains voltage to around 3 kV and the diode converts it into DC. The magnetron emits microwave energy that heats up food.

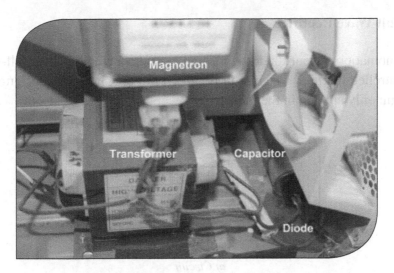

Fig.3.1 HV circuit of microwave oven. Caution: Around 3kV is produced by the HV transformer. Only qualified technicians can check the circuit.   © Mustafa Husain.

A single-phase, half-wave diode rectifier circuit is shown in Fig.3.2a. The ac signal is first stepped down from V_P to V_{in}, rectified by the diode and then applied across a purely resistive load. During the positive cycle of v_{in} (Fig.3.2b), the diode conducts and becomes an ideal switch because its anode is more positive than the cathode. Ideally, the voltage drop across the diode is zero but practically there is a voltage drop of around 0.7V for silicon diodes. The load voltage and current are in phase. During the negative cycle of v_{in} (Fig.3.2c), the anode becomes less positive than the cathode and thus the diode turns off and acts like an open switch. Therefore, the voltage across the diode is equal to that of v_{in}. The load voltage and current are zero. The output voltage waveform is shown in Fig.3.2d where it is made up of the positive wave of the input voltage. The average value of the output voltage is

$$V_{out} = 0.45 V_{in} \quad \textbf{Eq.3.1}$$

3.2 Full-Wave Rectifier

A combination of four diodes is formed to make a single-phase, full-wave uncontrolled rectifier. The circuit is shown in Fig.3.3a. This configuration is commonly applied in the industry.

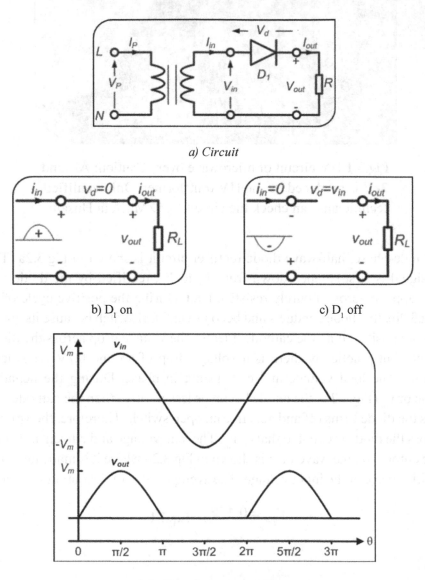

a) Circuit

b) D_1 on   c) D_1 off

d) Input and output waveforms

Fig.3.2 Half-wave uncontrolled rectifier.

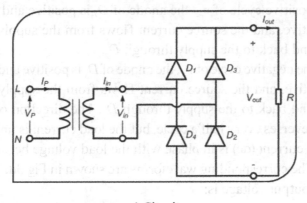

a) Circuit

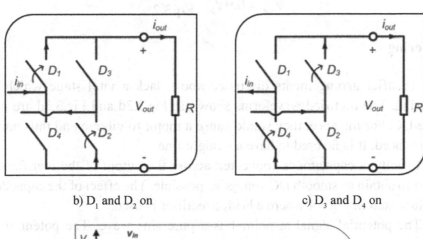

b) D_1 and D_2 on   c) D_3 and D_4 on

d) Input and output waveforms

Fig.3.3 Full-wave uncontrolled rectifier

During the positive cycle of v_{in}, the anode of D_1 is positive and the cathode of D_2 is negative, and the source current flows from the supply to the load through D_1 and back to the supply through D_2.

During the negative cycle of v_{in}, the anode of D_3 is positive and the cathode of D_4 is negative, and the source current flows from the supply to the load through D_3 and back to the supply through D_4. The direction of the source current (ac) reverses every half a cycle, but the load current is unidirectional.

The load current (dc) is in phase with the load voltage because the load is resistive. The corresponding waveforms are shown in Fig.3d. The average value of the output voltage is:

$$V_{out} = 0.9V_{in} \quad \textbf{Eq.3.2}$$

Filtering

The rectifier arrangements discussed above lack a vital stage which is filtering. The rectified waveforms shown in Fig.3.2d and Fig.3.3d are not pure DC. For instance, they would cause a motor to vibrate and thus noise is produced. It is favored to have a straight line.

Usually, a capacitor is connected across the output of the rectifier in order to obtain as smooth DC voltage as possible. The effect of the capacitor is illustrated in Fig.3.4 where a bridge rectifier is used.

The potential signal at point 1 is a pure sine wave. The potential at point 2 is a full wave. During the rising segment of this wave, the capacitor charges. As the wave tends to decrease, the capacitor discharges until another cycle appears where it charges again. The charging time is slower than the discharging time and so a smooth output voltage is produced.

With filtering, the output voltage increases beyond the values defined by Eq.3.1 and Eq.3.2 and up to the peak of the input voltage.

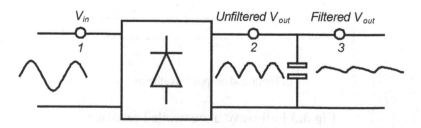

Fig.3.4 Filtering of rectified waveform

THYRISTOR RECTIFIERS

4.0 Introduction

Like the diode, the SCR can be used to rectify an AC signal. Unlike the diode, the conduction of an SCR can be controlled through its gate by external circuitry. This is known as controlled rectification because the output DC voltage can be varied by varying the triggering time (or angle) of the SCR.

4.1 Half-Wave Rectifier

A single-phase, half-wave controlled rectifier circuit is drawn in Fig.4.1a. The ac signal is first stepped down from V_P to V_{in}, rectified by the SCR and then applied across a purely resistive load.

During the positive cycle of v_{in} (Fig.4.1b), the anode is more positive than the cathode. The thyristor conducts and acts as a closed switch when a positive voltage pulse v_G is applied to its gate at an angle α.

During the negative cycle of v_{in} (Fig.4.1c), the anode becomes less positive than the cathode and thus the SCR turns off and acts like an open switch. The turn-off process of the thyristor here is called *natural line commutation*. Applying a positive pulse cannot turn on the SCR now.

The voltage and current waveforms are shown in Fig.4.1d where the output voltage is made up of the positive wave of the input voltage. The firing of the SCR is repeated after one cycle which is an angle of 360°. The average value of the load voltage is

$$V_{out} = 0.225(1 + \cos\alpha)V_{in} \quad \textbf{Eq.4.1}$$

A table showing the values of the output voltage based on different firing angles is helpful to understand this rectifier. The output voltage is high at very low firing angles. As the firing angle is increased, the output voltage decreases.

Table 4.1

α	V_{out}	Remarks
0	0.225 (1+1) V_{in} = 0.45 V_{in}	Just like an uncontrolled rectifier.
60°	0.225 (1+0.5) V_{in} = 0.3375 V_{in}	The output voltage is 33.75% of the input voltage
90°	0.225 (1+0) V_{in} = 0.225 V_{in}	The output voltage is 22.5% of the input voltage
120°	0.225 (1-0.5) V_{in} = 0.1125 V_{in}	The output voltage is 11.25% of the input voltage
180°	0.225 (1-1) V_{in} = 0	The output voltage vanishes

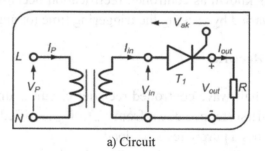

a) Circuit

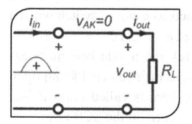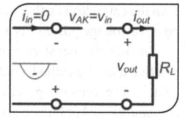

b) Operating modes.

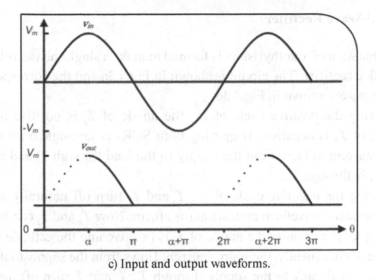

c) Input and output waveforms.

Fig.4.1 Single-phase, half-wave controlled rectifier.

Practical Circuit: A simple practical circuit to demonstrate the operation of a half-wave controlled rectifier can be constructed from basic components (Fig.4.2). The RC circuit produces the firing angle α which can be varied by varying the resistance R, and it can be determined from:

$$\alpha = \tan^{-1}\left(\frac{1}{2\pi f R C}\right) \quad \textbf{Eq.4.2}$$

where f is the frequency of the input voltage. The diode allows only positive pulses through the gate of the SCR. A DC filter is not included in the circuit but it can be added.

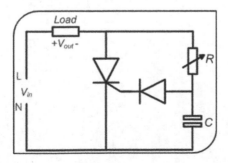

Fig.4.2 Practical circuit for realizing half-wave controlled rectifier.

4.2 Full-Wave Rectifier

A combination of four thyristors is formed to make a single-phase, full-wave controlled rectifier. The circuit is shown in Fig.4.3a and the corresponding waveforms are shown in Fig.4.3d.

During the positive cycle of v_{in}, the anode of T_1 is positive and the cathode of T_2 is negative. Triggering both SCRs at an angle α allows the source current to flow from the supply to the load through T_1 and back to the supply through T_2.

During the negative cycle of v_{in}, T_1 and T_2 turn off naturally and the whole negative waveform appears across them. Now T_3 and T_4 can be fired at an angle $\alpha+\pi$ because the anode of T_3 is positive and the cathode of T_4 is negative. Consequently, the source current flows from the supply to the load through T_3 and back to the supply through T_4. T_3 and T_4 turn off naturally during the positive cycle and the whole positive waveform appears across them.

The direction of the source current (AC) reverses every half a cycle, but the load current is unidirectional. The load current (DC) is in phase with the load voltage because the load is resistive. The firing of two SCRs together is repeated after an angle of 180°. The average value of the load voltage is

$$V_{out} = 0.45\,(1+\cos\alpha)V_{in} \quad \textbf{Eq.4.3}$$

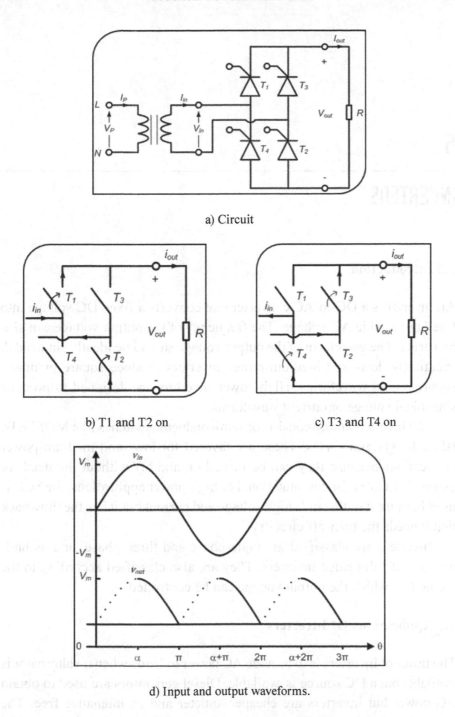

a) Circuit

b) T1 and T2 on

c) T3 and T4 on

d) Input and output waveforms.

Fig.4.3 Single-phase, full-wave controlled rectifier.

5

INVERTERS

5.0 Introduction

An inverter is a DC-to-AC converter that converts a fixed DC voltage into fixed or variable AC voltage. The frequency of the output voltage can also be varied. The waveform of the output voltage should be ideally sinusoidal. Practically, low- and medium-power inverters produce square or quasi-square output waveforms. High-power inverters are designed to produce sinusoidal voltage or current waveforms.

The inverter simply consists of semiconductor switches like MOSFETs, BJTs, IGBTs and GTOs. These are favored for low- and medium-power applications because they can be turned on and off without the need for external circuitry for commutation. For high-power applications, the SCR is used because it withstands high voltage and current but it has the drawback that it needs the turn-off circuitry.

Inverters are classified as single-phase and three-phase, and as half-bridge and full-bridge inverters. They are also classified according to the method by which the output voltage can be controlled.

5.1 Applications Of Inverters

The usage of inverters is to provide AC power to loads when no alternator is available but a DC source is available. Diesel generators are used to obtain AC power but inverters are cheaper, quieter and maintenance free. The inverter finds many applications. It can convert the DC voltage of solar cells into AC and hence it is possible to convert the solar energy into electrical energy which we need in our homes for lighting, air conditioning and so on. If you are having a picnic somewhere by the beach, you can use an inverter to convert the DC power of a spare car battery into AC power which you may need for a TV, lighting and a fan.

5.2 Half-Bridge Inverter

This inverter is fed with a battery as input, and it consists of two semiconductor switches. A typical circuit is shown in Fig.5.1a. A diode is connected in parallel across each switch. The diodes protect the switches against voltage spikes when they are off. The switches could be BJT, SCR, MOSFET, etc. In power applications, capacitors with very large and equal farads are employed to provide dc blocking through the load so that the load current does not have any dc component. The capacitors also overcome the problem of transformer saturation from the primary side if a transformer is used at the output to provide isolation.

Let us study the operation of the circuit in Fig.5.1a. One switch is turned on at a time. Let the period of the output signal be T, and let us assume each switch conducts for $T/2$. When S_1 is on and S_2 off, current flows from the positive terminal of the battery to S_1, through the load and C_2 and then back to the battery via the negative terminal as shown in Fig.5.1b. Half of the battery voltage appears across the load $(+V_{in}/2)$.

When S_2 is on and S_1 off, current flows from the positive terminal of the battery to C_1, through the load and S_2 and then back to the battery via the negative terminal (Fig.5.1c). Again, half of the battery voltage appears across the load $(-V_{in}/2)$. The waveform is shown in Fig.5.1d for a resistive load where the battery voltage is constant but the output voltage is a square wave that varies between $+V_{in}/2$ and $-V_{in}/2$. The load current is sinusoidal.

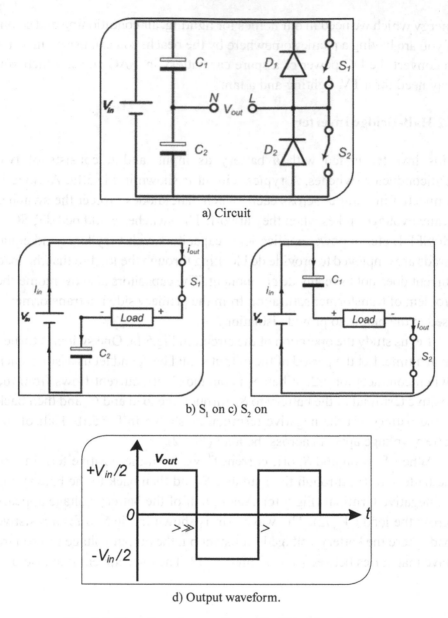

a) Circuit

b) S_1 on c) S_2 on

d) Output waveform.

Fig.5.1 Circuit and operation of half-bridge inverter.

The rms output voltage is

$$V_{out} = \frac{V_{in}}{2} \quad \textbf{Eq.5.1}$$

5.3 Bridge Inverter

A single-phase, full-bridge inverter employs four semiconductor switches. A bridge inverter with four IGBT's for photovoltaic applications is shown in Fig.5.2 followed by the power circuit. The control circuit is not included. The control circuit provides the control signals and these can be produced by electronic or microcontroller circuits.

The operation of the circuit is very simple. Only two switches are turned on at a time. When S_1 and S_2 are turned on, the current flows from the positive terminal of the battery to the load via S_1 and back to the battery via S_2 (Fig.5.4a).

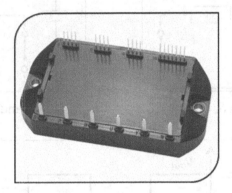

Fig.5.2 Bridge inverter module. © Powerex Inc.

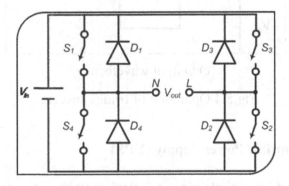

Fig.5.3 Bridge inverter circuit.

After a time $T/2$, S_1 and S_2 are turned off and S_3 and S_4 are turned on. The current flows from the positive terminal of the battery to S_3, through the load and then back to the battery via S_4 (Fig.5.4b). The full battery voltage appears across the load $(-V_{in})$.

The waveform is shown in Fig.5.4c for a resistive load where the output voltage is a square wave that varies between $+V_{in}$ and $-V_{in}$. The rms output voltage is

$$V_{out} = V_{in} \quad \textbf{Eq.5.2}$$

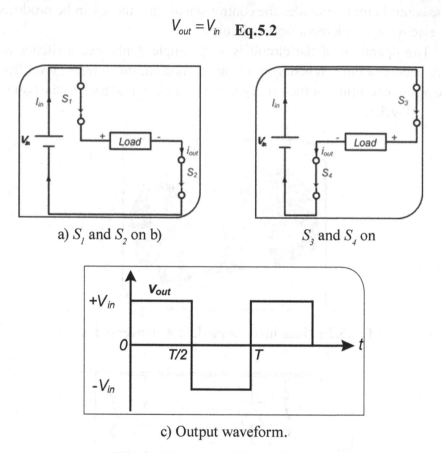

a) S_1 and S_2 on   b) S_3 and S_4 on

c) Output waveform.

Fig.5.4 Operation of bridge inverter.

5.4 Uninterruptible Power Supply (UPS)

The UPS is used in hospitals, airports, telephone companies and commercial malls to act as a stand by supply when the utility power is interrupted. Control centers at power stations use UPS for their computer systems and basic lighting. Hospitals need it for lighting and the operations of medical equipment especially in intensive care units and operations theatres.

Commercial malls need UPS systems for lighting, lifts and other important equipment. The applications of UPS are versatile and numerous. They play a big role in our life. The block diagram of a basic UPS system is shown in Fig.5.5. It consists of three main blocks:

- Charging system: a rectifier that charges a bank of battery often connected in series and parallel in order to have bigger current and voltage.
- Battery: it provides the power when utility power is interrupted
- Inverter: it converts the DC voltage of the battery into AC during power outages. The inverter incorporates a lot of filtering circuits so that sinusoidal ac voltage can be obtained at the output.
- The UPS has two inputs: ac input which is the power line, and dc input which is the battery. The output is ac voltage that is fed to the load. The UPS can be single-phase or three-phase.

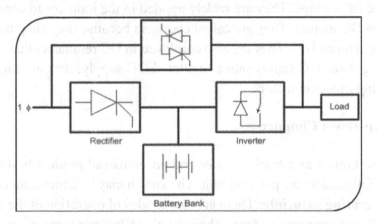

Fig.5.5 Block diagram of UPS system

6

CHOPPERS

6.0 Introduction

Choppers are DC-DC converters that convert a fixed DC voltage into a variable DC voltage. They are widely applied in the industry to control the speed of DC motors. They are called choppers because they chop the input DC signal on and off. They are also employed in DC regulators that convert an unregulated DC supply into a regulated DC supply; they are known as switching-mode regulators.

6.1 Step-Down Chopper

It is also known as a *buck chopper* and an industrial product is shown in Fig.6.1 followed by the power circuit. The switch may be a transistor or SCR.

Operating principle: There are two modes of operation of the switch. When the switch turns on for t_{on}, the inductive filter stores magnetic energy and its current, i_L, rises. When the switch is turned off, the stored energy in the inductor is released to the capacitor and load and the output current continues to flow. It is observed in Fig.6.3c that the input voltage is chopped and that the output current is DC. The buck chopper steps down DC voltage.

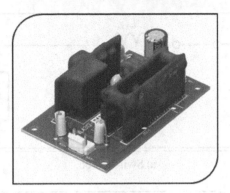

Fig.6.1 Step-down chopper. © Powerex Inc

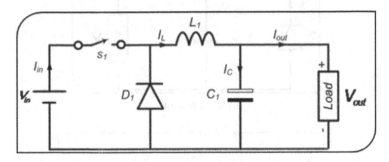

Fig.6.2 Buck chopper.

The average output voltage can be determined from:

$$V_o = kV_s \quad \textbf{Eq.6.1}$$

where k is the duty cycle defined as the ratio of the turn-on time t_{on} to the period T.

$$k = \frac{t_{on}}{T}; T = \frac{1}{f} \quad \textbf{Eq.6.2}$$

where f is the switching frequency of the semiconductor switch which is typically above 20 kHz.

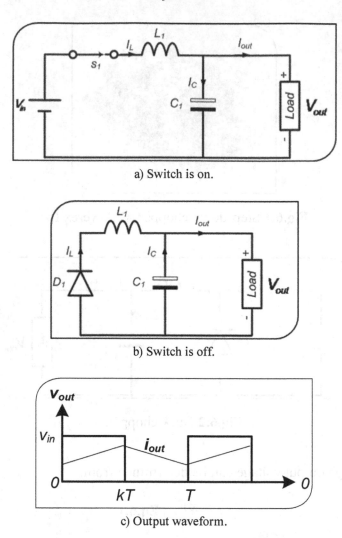

a) Switch is on.

b) Switch is off.

c) Output waveform.

Fig.6.3 Operating modes of step-down chopper.

6.2 Step-Up Chopper

It is also known as a step-up chopper and its circuit is depicted in Fig.6.4a. The boost chopper steps up DC voltage. The average output voltage can be determined from:

$$V_o = \frac{V_s}{1-k} \quad \textbf{Eq.6.3}$$

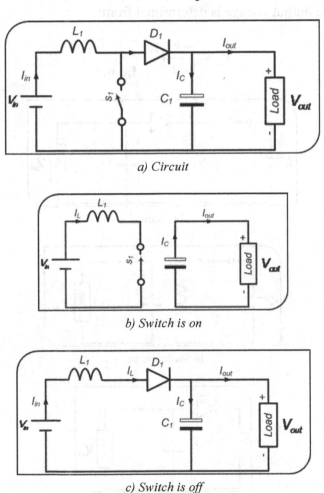

a) Circuit

b) Switch is on

c) Switch is off

Fig.6.4 Step-up chopper

6.3 Step Up/Down Chopper

In some applications, it is required to step up and step down dc voltage. A configuration was developed which integrates the operation of buck and boost choppers into one chopper known as buck-boost or step up-down chopper. This configuration is shown in Fig.6.5a.

This chopper steps up or down dc voltage depending on the duty cycle k. The average output voltage is determined from:

$$V_o = \frac{-k}{1-k} V_s \quad \textbf{Eq.6.4}$$

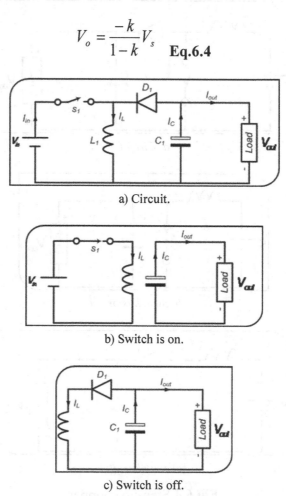

a) Circuit.

b) Switch is on.

c) Switch is off.

Fig.6.5 Step up/down chopper.

7

SINGLE-PHASE VOLTAGE REGULATORS

7.0 Introduction

AC voltage controllers are widely applied in industrial heating, speed control of induction motors, transformer tap-changing and light control (dimmers). Two SCRs connected inversely in parallel form a voltage controller. Hence, a TRIAC can also function as a voltage controller. However, the TRIAC is used for lower-power applications whereas the two SCRs can handle higher voltage and current than the TRIAC.

The power consumed by the load is controlled by two back-to-back SCRs (Fig.7.1). There are two methods of controlling the SCRs: phase-angle control and on-off control.

7.1 Phase-Angle Control

S_1 is triggered at an angle α during the positive cycle of v_{in} whereas S_2 is triggered at an angle $\pi+\alpha$ during the negative cycle. The waveforms are shown in Fig.7.2. Decreasing the angle α increases the power delivered to the source, and vice versa. It is possible to replace the SCRs with one TRIAC if it meets the power requirement. This load is often a lamp or a fan motor and the circuit is called a dimmer circuit.

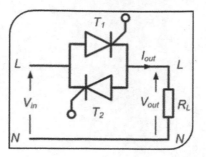

Fig.7.1 Single-phase voltage controller

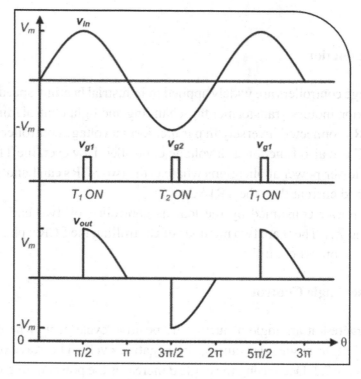

Fig.7.2 Phase-angle control method

7.2 On-Off Control

This is also referred to as integral half-cycle control method, and it employs the same circuit of Fig.7.1. The thyristors (or TRIAC) are triggered at the zero crossings of the AC input voltage. The input signal is passed on to the load for a number of integral half-cycles m, and it is interrupted for a number of integral half cycles n. The power supplied to the load can be controlled by controlling the ratio m/n. The waveforms are shown in Fig.7.3 for a ratio of 2/5. The rms output voltage is expressed as:

$$V_o = \sqrt{\frac{m}{m+n}}\, V_s = \sqrt{k}\, V_{in} \qquad \textbf{Eq.7.1}$$

where V_{in} is the AC input voltage and k is the duty cycle. From this equation we find for the output waveform in Fig.7.3 that 63% of the input power is supplied to the load. If it is desired to raise the output voltage, the number of on cycles must be increased.

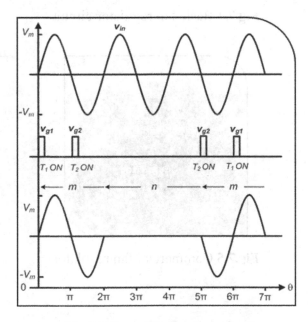

Fig.7.3 On-off control method

7.3 Commercial Application

Fan regulators and light dimmers are good applications of the phase-angle control method. Instead of two anti-parallel SCRs, a TRIAC is employed.

A typical circuit for a fan regulator is drawn in Fig.7.4 and an actual one is displayed thereafter. The ac filter is not included in the circuit diagram. The DIAC provides the pulses during both half cycles and the *RC* circuit decides the firing angle according to the value of the variable resistor.

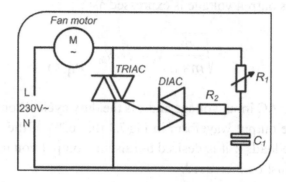

Fig.7.4 Basic fan regulator circuit

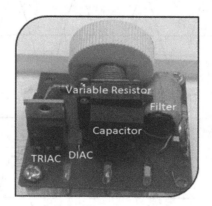

Fig.7.5 Commercial fan regulator

8

UNDERSTANDING DATA SHEETS

8.0 Introduction

Every electrical or electronic component, appliance, machine or system has voltage, current and power ratings. Power electronic devices are basically solid-state switches most of which have two ratings. One rating is for the control signal, and another for the power circuit. For instance, the SCR has a rating for the gate signal and another rating for the voltage across and current through the anode-cathode junction. The diode has only one rating because it is an uncontrolled device.

We will study how to select the best device for a certain application by understanding its data sheet. A device data sheet is published by the device manufacturer, and it contains detailed information about the device such as its rating, graphs, applications, etc. We have already studied the operation and characteristics of each device theoretically and practically, and so interpreting the information contained in a data sheet should be easy now.

8.1 The Diode

The diode rating is limited by the following ratings:

- Peak Inverse Voltage (PIV) or Peak Repetitive Reverse Voltage (V_{RRM}): The maximum value of the voltage the diode can withstand without being damaged when the diode is reverse biased i.e. when the cathode is positive and the anode is negative.

- Forward Voltage (V_F): The maximum voltage drop that occurs acorss the diode when the diode is ON.
- Forward Current (I_F): The rated average or RMS value of current that the diode can withstand when it is on.

8.2 The Thyristor

The thyristor here means either an SCR or a TRIAC. Both devices share the following definitions.

- Peak Repetitive off-State Forward Voltage (V_{DRM}) and Reverse Voltage (V_{RRM}): It is the maximum value of the voltage across the device when the device is off i.e. when no gate pulse applied. The device will remain off and undamaged unless these values are exceeded. Forward means when the anode (or MT_2) is positive and the cathode (or MT_1) is negative. Reverse means when the anode (or MT_2) is negative and the cathode (or MT_1) is positive.
- Peak Forward and Reverse Blocking Current (I_{DRM} & I_{RRM}): This is the maximum value of current the device can block when V_{DRM} or V_{RRM} is applied between the anode and cathode (or MT_2 and MT_1) with the gate left open. These values of current are very small leakage currents that flow through the device but do not mean the device is ON. Typical values are in micro amperes (μA).
- Peak On-State Voltage (V_{TM}): It is the voltage drop across the device when it is ON.
- On-State Current (I_{TM}): It is the RMS current that flows through the device when it is on. Often, the average value of current is also given. The values I_{TM} and V_{DRM} and V_{RRM} are the current and voltage ratings of the thyristor. Any technician or engineer looking for a thyristor should ask about these values first. In fact, these values are shown explicitly at the start of a data sheet.
- Holding Current (I_H): When the current flowing through the SCR between the anode and cathode drops below the holding current, the SCR turns off. This current is also defined for the TRIAC between MT_2 and MT_1.

- Gate Trigger Current: It is the gate current that turns on the thyristor. Its mimimum and maximum values are given in the data sheet, and the maximum value must not be exceeded.
- Gate Trigger Voltage: It is the voltage applied between the gate and cathode with the cathode being the reference. The minimum value is often 0.7V, and the maximum value must not be exceeded.

8.3 The BJT

In power electronics, the BJT is used as a switch. Suppose we want to select a transistor to switch a 12V, 2A incandescent lamp on and off as depicted in Fig.8.1. When the switch S is open (no base current), the transistor is OFF and so the full supply voltage of 12V appears across the transistor.

We should select a transistor that withstands the supply voltage. The voltage across the collector and emitter when the transistor is OFF is designated V_{CEO}. Therefore, we need to look for a transistor with $V_{CEO} >$ 12V. When the switch is closed, the transistor turns ON when enough base current flows. Then, the lamp turns on and its current is the collector current I_c. The resistor R should be calculated so that only enough base current flows.

Moreover, we should select a transistor with I_c rating of more than 2A. A voltage drop occurs between the collector and emitter due to the flow of the collector current, and so the transistor heats up. The voltage drop is designated $V_{CE(sat)}$ and it should be small so that almost all of the supply voltage appears across the lamp which means full lamp intensity. Because the switching speed here is not critical, we can ignore the turn-on time and turn-off time.

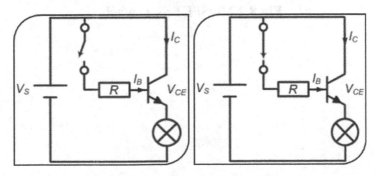

Fig.8.1 NPN transistor as a switch.

We should select a transistor that is a little overrated than 12V and 2A. A good choice may be selecting a 15V, 3A transistor.

8.4 The MOSFET

The MOSFET is mostly used in high-frequency applications due to its fast switching capability. However, let us again suppose we want to use it as a switch for the lamp in Fig.8.2. We should select a MOSFET with the following ratings:

- Gate to source voltage $V_{GS} > 12V$
- Drain current $I_D > 2A$

8.5 The IGBT

The IGBT can be selected just like selecting the BJT except that the IGBT has a voltage-controlled gate instead of a current-controlled base for the BJT. Suppose the IGBT is to be used instead of the BJT of Fig.8.1, then we should select 15V, 3A IGBT with gate voltage capability V_{GE} of 15V.

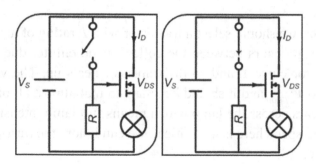

Fig.8.2 MOSFET as a switch.

PART II

EXPERIMENTS

TESTING OF POWER ELECTRONIC DEVICES

1.0 Objectives

- To learn how to test power electronic devices with a digital multimeter, and by using a lamp and a dc voltage supply

1.1 Equipment

Serial	Component/apparatus	Quantity
1	Breadboard	1
2	Power resistor 470Ω, 17W	1
3	DMM	2
4	12V lamp	1
5	1N5401 diode	1
6	2N6397 SCR	1
7	2N6073 TRIAC	1
8	2N3055 NPN transistor	1
9	IRL3303 N-type MOSFET	1
10	BUP400 IGBT	1
11	30V D.C. power supply	1

1.2 Discussion

Most of power semiconductor devices can be tested out of circuit with a DMM. The pn-junction silicon diode is the basis for understanding and applying DMM testing. Most of these devices have diode equivalent circuits. Therefore, the diode will be used for demonstration purposes.

There are two ways to test the diode with the DMM. The simplest way is to set the DMM to the diode mode. Here, the DMM supplies a constant current of 20 mA if the diode is connected like the demonstration presented in Fig.1.1.

Another way is to set the DMM to the ohm range as demonstrated in Fig.1.2. When the diode is forward biased, the ohmmeter measures the diode on-state resistance. If reverse biased, a healthy diode must exhibit an open circuit.

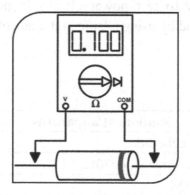

Fig.1.1 Diode testing

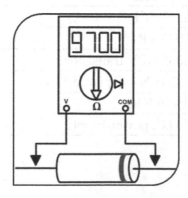

Fig.1.2 Ohm testing

Testing with the DMM is not always conclusive and it may lead to incorrect judgment whether the device is defective or not. As an alternative, a lamp and a dc source should be used to check the correct operational condition of the device. In this exercise, you will test a variety of devices.

1.3 Testing of Semiconductor Switches

1.3.1 Diode

1N5401 silicon diode will be used. Its fundamental ratings are listed in Table 1. These values must not be exceeded. The diode will be tested with the DMM, and forward and reverse operation will be conducted for more reliable judgment on its operational condition.

Table 1.1

Maximum rectified current	3A
Maximum reverse voltage	100V

Procedure

1. Set the DMM to the diode range. Test the diode with the DMM and record the results in Table 1.2.
2. With the power supply off and the voltage knob to the minimum setting, connect the circuit of Fig.1.3.
3. Turn on the power supply and set the voltage to 12Vdc.
4. Measure the voltage across the supply, diode and lamp. Record your measurements in Table 1.3.
5. Measure the lamp current and record the reading in Table 1.3. Switch off the power supply.
6. Reverse the connection of the diode as illustrated in Fig.1.4.
7. Turn on the supply. Repeat steps 4-6. You may need to set the ammeter range to μA. Record your readings in Table 1.3.
8. Write down a conclusion explaining what you have learned.

_____

_____

Table 1.2

DMM reading across anode-cathode	
DMM reading across cathode-anode	

Table 1.3	Forward (Fig.1.3)	Reverse (Fig.1.4)
Voltage across supply, V_{in}		
Voltage drop across diode, V_{AK}		
Voltage drop across lamp, V_{out}		
Lamp current, I_{AK}		

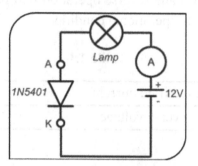

Fig.1.3 Circuit for testing diode in forward state.

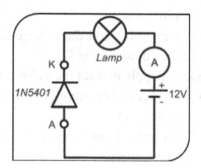

Fig.1.4 Circuit for testing diode in reverse state

1.3.2 SCR

2N6397 SCR will be tested in forward and reverse states. The voltage and current ratings are included in Table 1.4.

Table 1.4

Maximum rms on-state current	12 A
Maximum forward/reverse blocking voltage	400 V
Maximum gate trigger voltage	1.5 V
Maximum gate trigger current	30 mA
Typical holding current	6 mA

Procedure

1. Test the SCR with the DMM and record the results in Table 1.5.

Table 1.5

DMM reading across gate-cathode	
DMM reading across cathode-gate	
DMM reading across gate-anode	
DMM reading across cathode-gate	
DMM reading across anode-cathode	
DMM reading across cathode-anode	

2. Keeping the power supply off and the voltage knob to the minimum setting, connect the circuit of Fig.1.5.
3. Turn on the power supply and set the voltage to 12Vdc.

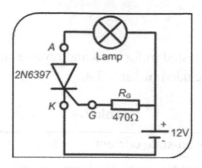

Fig.1.5 Circuit for testing SCR in forward state.

4. Measure the voltage across the supply, SCR and lamp. Record your measurements in Table 1.6.
5. Verify Kirchoff's voltage law by comparing the supply voltage with the sum of the voltage across the lamp and the voltage across the SCR.

    _____

    _____

6. Switch off the power supply. Make proper arrangements to measure the lamp current and the gate current simultaneously.

Table 1.6

Operating modes	Forward (Fig.1.5)	Reverse (Fig.1.6)
Voltage across supply, V_{in}		
Voltage drop across SCR, V_{AK}		
Voltage drop across lamp, V_{out}		
Lamp current with gate resistor connected		
Lamp current with gate resistor removed		
SCR gate current		

7. Turn on the supply and observe the DMM readings. Record the readings in Table 1.6.

8. Keeping the supply on and the DMMs connected as ammeters, remove the gate resistor. Observe the DMM readings. Record the readings in Table 1.6.

9. Switch off the power supply.

10. Reverse the connection of the SCR (Fig.1.6).

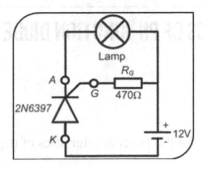

Fig.1.6 Circuit for testing SCR in reverse condition

11. Turn on the supply. Repeat steps 4-6. Record the readings in Table 1.6.

12. Write a conclusion explaining briefly what you have learned.

_____

_____

_____

_____

_____

EXPERIMENT 2

CHARACTERISTICS OF PN-JUNCTION DIODE

2.0 Objectives

- To study and plot the *V-I* characteristics of the diode

2.1 Equipment

Serial	Component/apparatus	Quantity
1	Breadboard	1
2	12V lamp, 20W	1
3	DMM	2
4	1N5401 diode	1
5	30V D.C. power supply	1

2.2 Discussion

If the relationship between the voltage across and the current through the pn-junction diode is obtained, the behavior and operation of the diode can be explored. More information about electrical quantities can be gathered such as the rated voltage, current, power and resistance.

This experiment will allow you to observe and record the behavior of the current through the diode when the voltage across its terminals is varied during two states: when the diode is forward biased and when it is reverse biased.

2.3 Experiment

2.3.0 Forward Bias

Procedure

1. Turn on the power supply and set the output to 0V. Switch off the supply.
2. Connect the circuit of Fig.2.1. Set the ammeter to the mA range.

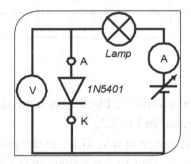

Fig.2.1 Forward-biased diode

3. Turn on the supply. Vary the output of the supply so that the voltage across the diode V_{AK} is 0.1V.
4. Measure and record the current in Table 2.1.
5. Repeat step 4 by increasing V_{AK} in steps of 0.1V until 1V. Record the corresponding values of I_{AK}.
6. Reset the output of the supply to 0 V. Switch off the supply.
7. For each row in Table 2.1, calculate the ratio V_{AK} / I_{AK} and put the results in the table.

Table 2.1

V_{AK} (V)	I_{AK} (mA)	V_{AK} / I_{AK}
0		
0.10		
0.20		
0.30		
0.40		
0.50		
0.60		
0.70		
0.80		
0.90		
1.00		

2.3.1 Reverse Bias

1. Using the same circuit of Fig.2.1, reverse the connections of the diode as illustrated in Fig.2.2.
2. Set the ammeter range to μA. Turn on the power supply. Gradually increase V_{AK} according to the values listed in Table 2.2. Measure and record the corresponding values of I_{AK}.
3. Calculate the ratio V_{AK} / I_{AK} and put the results in the Table 2.2.

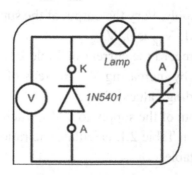

Fig.2.2 Reverse-biased diode

Table 2.2

$V_{AK}(V)$	$I_{AK}(\mu A)$	V_{AK}/I_{AK}
0		
-1.00		
-2.00		
-3.00		
-4.00		
-5.00		
-6.00		
-7.00		
-8.00		
-9.00		
-10.00		
-11.00		
-12.00		
-13.00		
-14.00		
-15.00		
-16.00		
-17.00		
-18.00		
-19.00		
-20.00		
-21.00		
-22.00		
-23.00		
-24.00		
-25.00		

4. Plot the results of Table 2.1 and Table 2.2 in Fig.2.3. Put the values of I_{AK} in the x-axis and the values of V_{AK} in the y-axis.

5. Summarize your conclusion

_____

_____

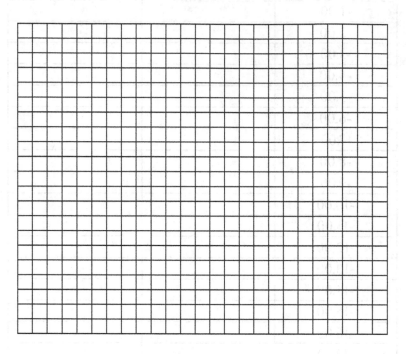

Fig.2.3 V-I curve

EXPERIMENT 3

UNCONTROLLED AC-DC CONVERTER

3.0 Objectives

- To study the operation of the diode as a half-wave rectifier
- To study the operation of the diode as a full-wave rectifier

3.1 Equipment

Serial	Component/apparatus	Quantity
1	Breadboard	1
2	Power resistor 680Ω, 17W	1
3	12V dc motor	1
4	DMM	2
5	1N5401 diode	4
6	DF06 bridge rectifier IC	1
7	15-0-15V ac power supply	1

3.2 Discussion

The conversion of ac voltage into dc voltage will be carried out in this experiment. The experiment will be conducted in three stages. In the first stage, you will study the operation of the diode as a half-wave rectifier. The same circuit will be used in the second stage but with adding three more diodes so that the four diodes operate as a full-wave rectifier. In the last stage, the four diodes will be replaced with one chip containing four diodes.

In all stages, the value and waveform of the output voltage will be obtained. Caution must be taken as a polarized capacitor will be used. The positive terminal of the capacitor must be always connected to a positive terminal.

3.3 Experiment

3.3.0 Half-Wave Rectifier

Procedure

1. With the power supply off, connect the circuit of Fig.3.1.

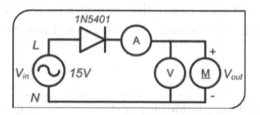

Fig.3.1 Half-wave rectifier

2. Turn on the supply.
3. Measure the input voltage, output voltage and load current. Put the results in Table 3.1.
4. Put your hand on the chassis of the motor. Write down your observation:

_____

5. Draw the waveforms of the input and output voltages in Fig.3.2.
6. Switch off the supply.
7. Calculate the ratio V_{out}/V_{in} and insert the result in Table 3.1.
8. Connect the capacitor across the motor. The positive terminal of the capacitor must be connected to the cathode of the diode and the negative terminal to the neutral of the ac supply.
9. Switch on the supply.

10. Put your hand on the chassis of the motor. Write down your observation:

_____

11. Measure the input voltage, output voltage and load current. Put the results in Table 3.1.

Table 3.1

Quantity	Measured value without capacitor	V_{out}/V_{in}	Measured value with capacitor	V_{out}/V_{in}
$V_{in}(V)$				
$V_{AK}(V)$				
$V_{out}(V)$				
$I_{AK}(mA)$				

12. Draw the waveforms of the input and output voltages in Fig.3.2 and Fig.3.3.
13. Switch off the supply.
14. Calculate the ratio V_{out}/V_{in} and insert the result in Table 3.1.
15. Summarize what you have learned:

_____

_____

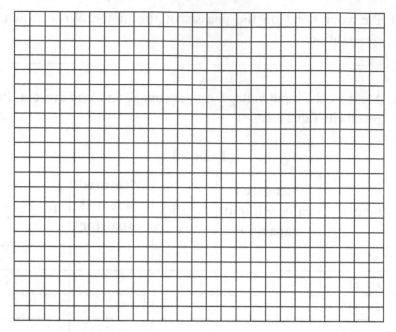

Fig.3.2 Voltages without a capacitor

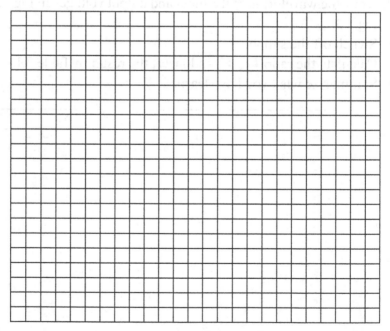

Fig.3.3 Voltages with a capacitor

3.3.1 Full-Wave Rectifier Using Discrete Diodes

Procedure

1. With the power supply off, connect the circuit of Fig.3.4.

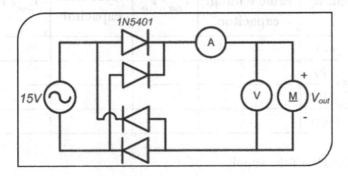

Fig.3.4 Full-wave rectifier

2. Turn on the supply.
3. Measure the input voltage, output voltage and load current. Put the results in Table 3.2.
4. Put your hand on the chassis of the motor. Write down your observation compared with the situation of using one diode:

_____

_____

5. Draw the waveforms of the input and output voltages in Fig.3.5.
6. Switch off the supply.
7. Calculate the ratio V_{out}/V_{in} and insert the result in Table 3.2.
8. Connect the capacitor across the motor minding the polarity.
9. Switch on the supply.
10. Measure the input voltage, output voltage and load current. Put the results in Table 3.2.
11. Put your hand on the chassis of the motor. Write down your observation:

_____

12. Draw the waveforms of the input and output voltages in Fig.3.6.

Table 3.2

Quantity	Measured value without capacitor	V_{out}/V_{in}	Measured value with capacitor	V_{out}/V_{in}
$V_{in}(V)$				
$V_{AK}(V)$				
$V_{out}(V)$				
$I_{AK}(mA)$				

13. Switch off the supply.
14. Calculate the ratio V_{out}/V_{in} and insert the result in Table 3.2.
15. Summarize what you have learned:

_____

_____

_____

_____

_____

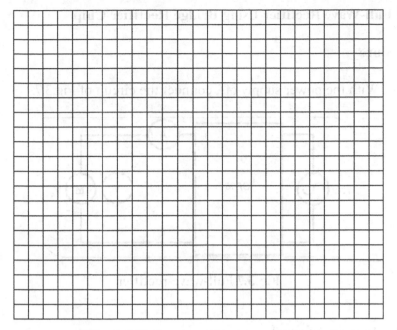

Fig.3.5 Voltages without a capacitor

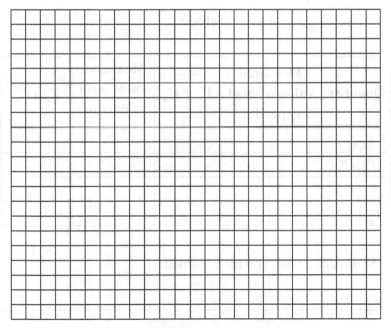

Fig.3.6 Voltages with a capacitor

3.3.2 Full-Wave Rectifier Using Bridge Rectifier Chip

Procedure

1. With the power supply off, connect the circuit of Fig.3.7.

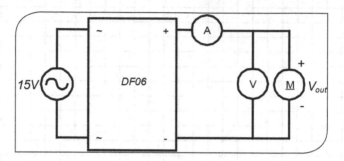

Fig.3.7 Full-wave rectifier

2. Turn on the supply.
3. Fill in Table 3.3 using the same procedure of section 3.3.1.

Table 3.3

Quantity	Measured value without capacitor	V_{out}/V_{in}	Measured value with capacitor	V_{out}/V_{in}
$V_{in}(V)$				
$V_{AK}(V)$				
$V_{out}(V)$				
$I_{AK}(mA)$				

4. Summarize what you have learned:

_____

_____

_____

EXPERIMENT 4

CONTROLLED AC-DC CONVERTER

4.0 Objectives

- To study the operation of the SCR as a half-wave rectifier
- To study the operation of the SCR as a full-wave rectifier

4.1 Equipment

Serial	Component/apparatus	Quantity
1	Breadboard	1
2	Power resistor 680Ω, 17W	1
3	12V dc motor	1
4	DMM	2
5	2N6397 SCR	4
6	DF06 bridge rectifier IC	1
7	15-0-15V ac power supply	1

4.2 Discussion

The conversion of ac voltage to fixed dc voltage was conducted in the previous experiment using the pn-junction diode. In this experiment, you will be able to vary the dc voltage using the SCR instead of the diode. A half-wave circuit will be used.

The SCR will be triggered with an RC circuit, and the firing angle will be changed gradually from 0° to 180° by a variable resistor. The capacitor of the control circuit is non-polarized. The filter capacitor is polarized and so the polarity must be known and connected correctly.

The load is a dc motor so that you can see the effect of variable voltage on motor speed.

4.3 Experiment

Procedure

1. With the power supply off, connect the circuit of Fig.4.1.

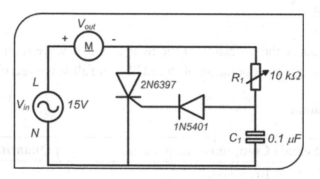

Fig.4.1 Half-wave controlled rectifier

2. Set the variable resistor, R_1, to the middle position.
3. Turn on the supply.
4. Measure the input voltage, output voltage, motor current and gate current. Put the results in Table 4.1.
5. Connect Channel 1 of the CRO across the ac supply and Channel 2 across the motor. The positive terminal must be connected to the line and the negative terminal to the neutral. Draw the input waveform in Fig.4.1.
6. Connect Channel 2 of the CRO across the motor. Draw the output waveforms in Fig.4.2.
7. While gradually reducing the value of R_1, observe the motor speed and the output waveform.

8. Set R_l to its minimum value. Measure the input voltage, output voltage, motor current and gate current. Put the results in Table 4.1.
9. Plot the output waveform in Fig.4.2 just below the previous waveform.
10. Set R_l to its maximum value. Measure the input voltage, output voltage, motor current and gate current. Put the results in Table 4.1.
11. Plot the output waveform in Fig.4.2 just below the previous waveforms.
12. Summarize what you have learned:

_____

_____

_____

Table 4.1

Quantity	Measured value		
	R_l = mid	R_l = min	R_l = max
Input voltage, $V_{in}(V)$			
Output voltage, $V_{out}(V)$			
Motor current, $I_{AK}(mA)$			
Gate current, $I_G(mA)$			

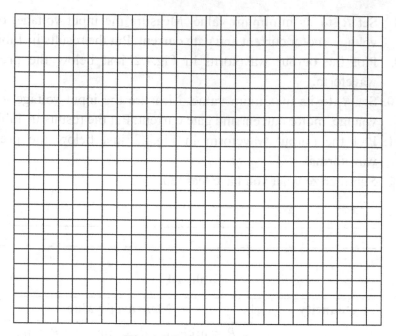

Fig.4.2 Input waveform

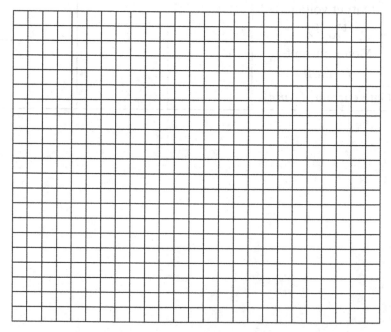

Fig.4.3 Output waveform

EXPERIMENT 5

DC-DC CONVERTER

5.0 Objectives

- To learn the basic concept of DC-DC converter
- To study the effect of duty cycle on the output voltage

5.1 Equipment

Serial	Component/apparatus	Quantity
1	Resistor 100Ω, 17W	1
2	Resistor 1kΩ	1
3	Inductor 15 mH	1
4	Polarized Capacitor 1000 µF	1
5	TIP31A NPN transistor	1
6	1N5401 diode	1
7	DC power supply	1
8	Function generator	1
9	Oscilloscope	1

5.2 Discussion

In this experiment, you will build a DC-DC converter circuit that can step down DC input voltage. The circuit is composed of two basic elements: two semiconductor switches and two filters. A BJT will be used as the controlled switch and a pn-junction diode as the uncontrolled switch. A fast recovery diode is highly recommended for proper operation of the converter.

With a constant-frequency control signal supplied from a function generator, the circuit can vary the output voltage by varying the turn-on time of the BJT.

5.3 Experiment

5.3.0 Step-Down Chopper

Procedure

1. Connect the main output of the function generator to Channel 1 of the CRO. Turn on the function generator and the CRO.
2. Adjust the main output of the function generator to produce a square wave with a peak-to-peak voltage of 20V at 1 kHz.
3. Keep the mark-space knob of the function generator off.
4. Turn off the function generator and CRO.
5. Turn on the power supply and adjust it to produce 15V dc.
6. Turn off the supply.
7. Connect the circuit of Fig.5.1.

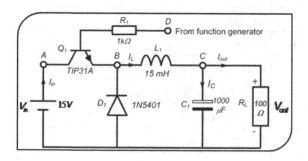

Fig.5.1 Step-down chopper

8. Connect Channel 2 of the CRO across the load.
9. Connect the output of the function generator to the base of the transistor via the resistor R_1.
10. Turn on the supply and CRO.

11. Measure the voltages at points *A*, *B* and *C* with respect to the negative terminal of the supply. Record the results in Table 5.1.
12. Turn on the function generator. Repeat step 11.

Table 5.1

Quantity	Measured value without control signal	Measured value with control signal
V_A (V)		
V_B (V)		
V_C (V)		

13. Plot the waveform of the load voltage in Fig.5.2.
14. Remove the capacitor.
15. Plot the waveform of the load voltage in Fig.5.3.
16. Turn off the supply, CRO and function generator.
17. Summarize what you have learned:

_____

_____

_____

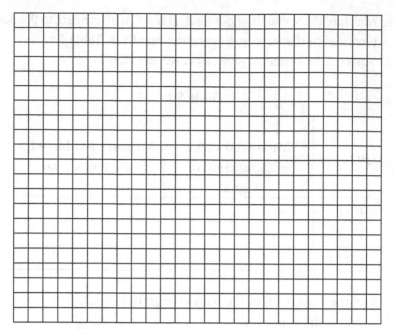

Fig.5.2 Output waveform with capacitor

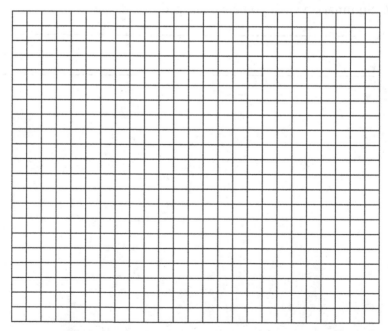

Fig.5.3 Output waveform without capacitor

EXPERIMENT 6

AC-AC CONVERTER

6.0 Objectives

- To learn the basic concept of ac-ac converter
- To study the phase-angle and on-off method

6.1 Equipment

Serial	Component/apparatus	Quantity
1	15V ac power supply	1
2	2N6073 TRIAC	1
3	1N5401 diode	2
4	Resistor 330Ω	1
5	Variable resistor 10 kΩ	1
6	Non-polarized capacitor 100 μF	1
7	4N35 and MOC3011 optocouplers	1
8	PIC16F628	1

83

6.2 Discussion

The lamp dimmer is a good application of AC-AC converters which convert a fixed AC voltage into variable values. You will experiment with two methods.

In the first part of the experiment, you will build a simple lamp dimmer using two diodes, one TRIAC and some basic components; this is the phase-angle method. You will vary an input voltage of 15V~ by varying the firing angle with a variable resistor and you will observe this change on the voltmeter and the CRO.

The second part is the on-off control method. You will use a microcontroller, two SCRs and an optocoupler. The recommended microcontroller is the well-known and cheap PIC16F628.

6.3 Experiment

6.3.0 Phase-Angle Control Method

Procedure

1. With the power supply off, connect the circuit of Fig.6.1.
2. Set the variable resistor to roughly its minimum position.
3. Connect Channel 1 of the CRO across the lamp and Channel 2 between the TRIAC gate terminal and the neutral of the supply.
4. Connect a voltmeter across the lamp and connect an ammeter in series with the TRIAC gate terminal.
5. Turn on the supply and the CRO.
6. Observe the waveforms of the lamp voltage and the gate voltage on the CRO.

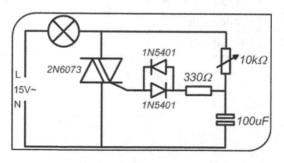

Fig.6.1 Dimmer circuit

Table 6.1

Quantity	Measured value		
	R_1 = mid	R_1 = min	R_1 = max
Output voltage, V_{out} (V)			
Gate current, I_G (mA)			

7. Record the voltage and current readings in Table 6.1.
8. Plot the waveforms in Fig.6.2.
9. Set the variable resistor to roughly its middle position.
10. Record the voltage and current readings in Table 6.1.
11. Plot the waveforms in Fig.6.3.
12. Set the variable resistor to roughly its maximum position.
13. Record the voltage current readings in Table 6.1.
14. Plot the waveforms in Fig.6.4.
15. Turn off the power supply and CRO.
16. Write down your conclusion

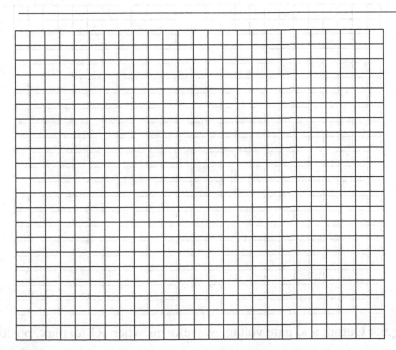

Fig.6.2 Output and gate voltage waveforms when R1 @ min. position

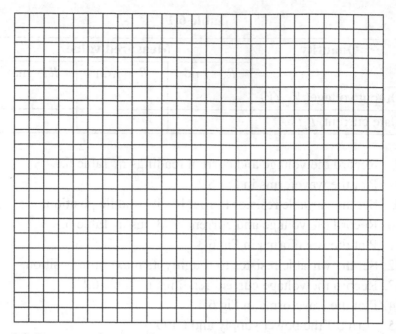

Fig.6.3 Output and gate voltage waveforms when R1 @ mid. Position

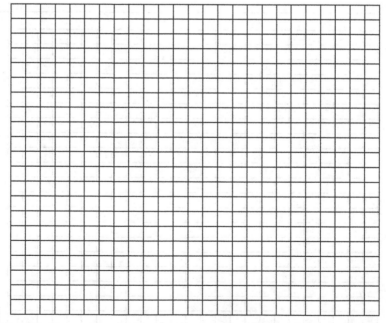

Fig.6.4 Output and gate voltage waveforms when R1 @ max. position

6.3.1 On-Off Control Method

Pre-start

In this part, the RC circuit and the diodes will be removed. The triggering pulses will be provided by the microcontroller PIC16F628A. The input signal will be isolated from the triggering circuit by the optocoupler MOC2021.

Prior to starting, the code for generating the triggering pulses must be loaded on the PIC16F628A. The code is listed in Table 6.2.

Table 6.2

```
INCLUDE<P16F628A.INC>

#DEFINE      PULSE       PORTB,2      ;crossing zero pin

#DEFINE      OUTPUT      PORTB,4      ;output from pic to optocup

#DEFINE      UP          PORTB,0      ;push button to change setup

SPEED        EQU         H'20'        ;memory for the on cycles running loop

SPEED1       EQU         H'21'        ;memory for the off cycles running loop

ON           EQU         H'22'        ;memory for the on cycles setup

OFF          EQU         H'23'        ;memory for the off cycles setup

DEL          EQU         H'24'        ;memory for the case setup

COUNT1       EQU         H'25'        ;memory for the delay loop

        BANKSEL     TRISB

        MOVLW       B'00001111'

        MOVWF       TRISB            ; set portb0 to portb3 input the rest output

        BANKSEL     PORTB

        CLRF        PORTB            ; make sure to reset all the memory

        CLRF        SPEED

        CLRF        SPEED1

        CLRF        ON

        CLRF        OFF
```

```
TRICK

        CLRF        DEL             ; this is for the reset case which is last one
START

        BTFSC       UP              ;off case and testing for the next case
        GOTO        INCREASE
        GOTO        START
RUN                                           ; run subroutine
        MOVF        ON,W
        MOVWF       SPEED
LOOP1                               ;this is the loop for the trigger pulse(on cycles)
        CALL        PULSETEST
        CALL        TRIG
        DECFSZ      SPEED
        GOTO        LOOP1
        MOVF        OFF,W
        MOVWF       SPEED1
LOOP2                               ;this is the loop for the off cycles
        CALL        PULSETEST
        BCF         OUTPUT
        DECFSZ      SPEED1
        GOTO        LOOP2
        CALL        TEST
        GOTO        SETUP
```

```
INCREASE                              ;change to next setup subroutine

        CALL      HOLD

        INCF      DEL

        GOTO      SETUP

SETUP                                 ;selection of the cases subroutine

        MOVF      DEL,0

        ADDWF     PCL,1

        NOP

        GOTO      CASE1

        GOTO      CASE2

        GOTO      CASE3

        GOTO      CASE4

        GOTO      CASE5

CASE1                         ;1 on and 15 off

        MOVLW     D'1'

        MOVWF     ON

        MOVLW     D'15'

        MOVWF     OFF

        GOTO      RUN

CASE2                         ;1 on and 4 off

        MOVLW     D'1'

        MOVWF     ON

        MOVLW     D'4'

        MOVWF     OFF

        GOTO   RUN
```

```
CASE3                                    ;2 on and 2 off
        MOVLW       D'2'
        MOVWF       ON
        MOVLW       D'2'
        MOVWF       OFF
        GOTO   RUN
CASE4                                    ;9 on and 7 off
        MOVLW       D'9'
        MOVWF       ON
        MOVLW       D'7'
        MOVWF       OFF
        GOTO        RUN
CASE5
        GOTO        TRICK        ; off case
HOLD                             ;making sure that you release the push button
        BTFSC       UP
        GOTO        HOLD
        RETURN
TEST                             ;testing the push button
        BTFSC       UP
        GOTO        INCREASE
        RETURN
```

```
PULSETEST                           ;testing the pulse and make sure that you get one pulse each time

        BTFSS        PULSE

        GOTO         PULSETEST

P1      BTFSS        PULSE

        RETURN

        GOTO         P1

TRIG                                ;trigger pulse for the output

        BSF          OUTPUT

        CALL         DELAY

        BCF          OUTPUT

        RETURN

DELAY                               ; delay used by the trigger subroutine( pulse width of triac)

        movlw        d'120'

        movwf        COUNT1

loop1P

        decfsz       COUNT1

        goto         loop1P

        RETURN

END
```

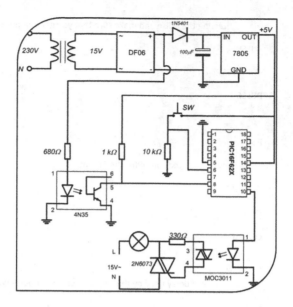

Fig.6.5 Dimmer circuit using on-off method

Procedure

1. Load the code listed in Table 6.2 onto the PIC16F628A.
2. With the power supply off, connect the circuit of Fig.6.5.
3. Connect Channel 1 of the CRO across the lamp and Channel 2 between the TRIAC gate terminal and the neutral of the supply.
4. Connect a voltmeter across the lamp and connect an ammeter in series with the TRIAC gate terminal.
5. Turn on the supply and the CRO.
6. Press the switch SW once. Observe the waveforms of the lamp voltage and the gate voltage on the CRO.
7. Record the voltage and current readings in Table 6.3.
8. Plot the waveforms in Fig.6.6.
9. Press the switch SW again.
10. Record the voltage and current readings in Table 6.3.
11. Plot the waveforms in Fig.6.7.
12. Press the switch SW once more.
13. Record the voltage and current readings in Table 6.3.
14. Plot the waveforms in Fig.6.8.

15. Turn off the power supply and CRO.
16. Write down your conclusion

_____

_____

_____

Table 6.3

Quantity	Measured value		
	SW = first time pressed	SW = second time pressed	SW = third time pressed
Output voltage, Vout (V)			
Gate current, IG (mA)			

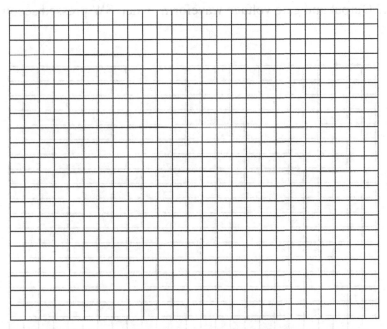

Fig.6.6 Output and gate voltage waveforms when SW pressed first time

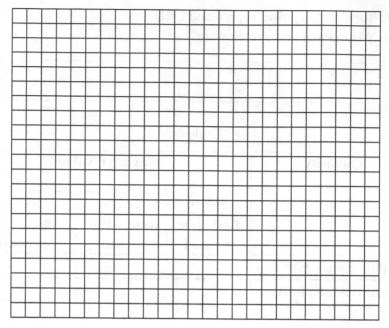

Fig.6.7 Output and gate voltage waveforms when SW pressed second time

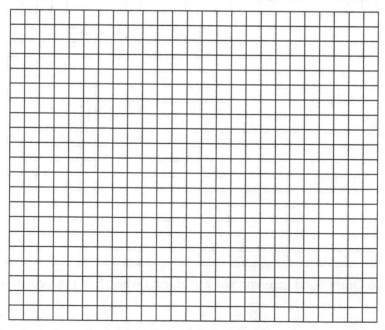

Fig.6.8 Output and gate voltage waveforms when SW pressed third time

DATA SHEETS

3.0A RECTIFIER

Features

f   Diffused Junction
f   High Current Capability and Low Forward Voltage Drop
f   Surge Overload Rating to 200A Peak
f   Low Reverse Leakage Current
f   **Lead Free Finish, RoHS Compliant (Note 3)**

Mechanical Data

f   Case: DO-201AD
f   Case Material: Molded Plastic. UL Flammability
    Classification Rating 94V-0
f   Moisture Sensitivity: Level 1 per J-STD-020C
f   Terminals: Finish — Tin. Plated Leads Solderable per
    MIL-STD-202, Method 208 ⓔ③
f   Polarity: Cathode Band
f   Marking: Type Number
f   Weight: 1.1 grams (approximate)

DO-201AD		
Dim	Min	Max
A	25.40	—
B	7.20	9.50
C	1.20	1.30
D	4.80	5.30
All Dimensions in mm		

Maximum Ratings and Electrical Characteristics @ T_A = 25°C unless otherwise specified

Single phase, half wave, 60Hz, resistive or inductive load.
For capacitive load, derate current by 20%.

Characteristic		Symbol	1N 5400	1N 5401	1N 5402	1N 5404	1N 5406	1N 5407	1N 5408	Unit
Peak Repetitive Reverse Voltage Working Peak Reverse Voltage DC Blocking Voltage		V_{RRM} V_{RWM} V_R	50	100	200	400	600	800	1000	V
RMS Reverse Voltage		$V_{R(RMS)}$	35	70	140	280	420	560	700	V
Average Rectified Output Current	@ T_A = 105°C (Note 1)	I_O	3.0							A
Non-Repetitive Peak Forward Surge Current 8.3ms Single half sine-wave superimposed on rated load		I_{FSM}	200							A
Forward Voltage	@ I_F = 3.0A	V_{FM}	1.0							V
Peak Reverse Current at Rated DC Blocking Voltage	@ T_A = 25°C @ T_A = 150°C	I_{RM}	10 100							μA
Typical Total Capacitance	(Note 2)	C_T	50				25			pF
Typical Thermal Resistance Junction to Ambient		$R_{\theta JA}$	15							°C/W
Operating and Storage Temperature Range		T_j, T_{STG}	-65 to +150							°C

Notes: 1. Valid provided that leads are kept at ambient temperature at a distance of 9.5mm from the case.
        2. Measured at 1.0MHz and applied reverse voltage of 4.0V DC.
        3. RoHS revision 13.2.2003. Glass and High Temperature Solder Exemptions Applied, see *EU Directive Annex Notes 5 and 7.*

1.0A SURFACE MOUNT GLASS PASSIVATED BRIDGE RECTIFIER

Features

- Glass Passivated Die Construction
- Low Forward Voltage Drop, High Current Capability
- Surge Overload Rating to 50A Peak
- Designed for Surface Mount Application
- UL Listed Under Recognized Component Index, File Number E94661
- **Lead-Free Finish; RoHS Compliant (Notes 1 & 2)**

Mechanical Data

- Case: DF-S
- Case Material: Molded Plastic. UL Flammability Classification Rating 94V-0
- Moisture Sensitivity: Level 1 per J-STD-020C
- Terminals: Finish - Tin. Solderable per MIL-STD-202, Method 208 ⓔ③
- Polarity: As Marked on Case
- Weight: 0.38 grams (approximate)

Ordering Information (Note 3)

Part Number	Case	Packaging
DFxS	DF-S	50 Per Tube
DFxS-T	DF-S	1500/Tape & Reel, 13-inch

Notes: 1. EU Directive 2002/95/EC (RoHS) & 2011/65/EU (RoHS 2) compliant. All applicable RoHS exemptions applied.
2. See http://www.diodes.com/quality/lead_free.html for more information about Diodes Incorporated's definitions of Halogen- and Antimony-free, "Green" and Lead-free.
3. For packaging details, go to our website at http://www.diodes.com/products/packages.html.

Marking Information

DFxxxS = Product Type Marking Code, ex: DF10S
YWW = Date Code Marking
Y = Last digit of year (ex: 2 for 2012)
WW = Week code 01 to 52

DF005S – DF10S
Document number: DS17001 Rev. 16 - 2

1 of 5
www.diodes.com

August 2013
© Diodes Incorporated

Maximum Ratings (@T$_A$ = +25°C, unless otherwise specified.) Single

phase, half wave, 60Hz, resistive or inductive load
For capacitive load, derate current by 20%.

Characteristic	Symbol	DF 005S	DF 01S	DF 02S	DF 04S	DF 06S	DF 08S	DF 10S	Unit
Peak Repetitive Reverse Voltage Working Peak Reverse Voltage DC Blocking Voltage	V$_{RMM}$ V$_{RWM}$ V$_R$	50	100	200	400	600	800	1000	V
RMS Reverse Voltage	V$_{RMS}$	35	70	140	280	420	560	700	V
Average Forward Rectified Current @ T$_A$ = +40°C	I$_O$				1.0				A
Non-Repetitive Peak Forward Surge Current, 8.3 ms Single Half Sine-Wave Superimposed on Rated Load	I$_{FSM}$				50				A
Non-Repetitive Peak Forward Surge Current, 1.0 ms Single Half Sine-Wave Superimposed on Rated Load	I$_{FSM}$				100				A

Thermal Characteristics

Characteristic	Symbol	DF 005S	DF 01S	DF 02S	DF 04S	DF 06S	DF 08S	DF 10S	Unit
Typical Thermal Resistance, Junction to Ambient (Note 2)	R$_{\theta JA}$				+40				°C/W
Operating and Storage Temperature Range	T$_J$, T$_{STG}$				-65 to +150				°C

Electrical Characteristics (@T$_A$ = +25°C, unless otherwise specified.)

Characteristic	Symbol	DF 005S	DF 01S	DF 02S	DF 04S	DF 06S	DF 08S	DF 10S	Unit
Forward Voltage (per element) @ I$_F$ = 1.0A	V$_{FM}$				1.1				V
Peak Reverse Current at Rated @ T$_A$ = +25°C DC Blocking Voltage (per element) @ T$_A$ = +125°C	I$_{RM}$				10 500				µA
I^{2}t Rating for Fusing (t<8.3ms)	I^{2}t				10.4				A^{2}s
Typical Total Capacitance (per element) (Note 1)	C$_T$				25				pF

2N6394 Series

Silicon Controlled Rectifiers
Reverse Blocking Thyristors

Designed primarily for half-wave ac control applications, such as motor controls, heating controls and power supplies.

Features
- Glass Passivated Junctions with Center Gate Geometry for Greater Parameter Uniformity and Stability
- Small, Rugged, Thermowatt Construction for Low Thermal Resistance, High Heat Dissipation and Durability
- Blocking Voltage to 800 V
- These are Pb−Free Devices

MAXIMUM RATINGS † (T_J = 25°C unless otherwise noted)

Rating	Symbol	Value	Unit
Peak Repetitive Off−State Voltage (Note 1) (T_J = −40 to 125°C, Sine Wave, 50 to 60 Hz, Gate Open) 2N6394 2N6395 2N6397 2N6399	V_{DRM}, V_{RRM}	50 100 400 800	V
On−State RMS Current (180° Conduction Angles; T_C = 90°C)	$I_{T(RMS)}$	12	A
Peak Non−Repetitive Surge Current (1/2 Cycle, Sine Wave, 60 Hz, T_J = 90°C)	I_{TSM}	100	A
Circuit Fusing (t = 8.3 ms)	I^2t	40	A²s
Forward Peak Gate Power (Pulse Width ≤ 1.0 μs, T_C = 90°C)	P_{GM}	20	W
Forward Average Gate Power (t = 8.3 ms, T_C = 90°C)	$P_{G(AV)}$	0.5	W
Forward Peak Gate Current (Pulse Width ≤ 1.0 μs, T_C = 90°C)	I_{GM}	2.0	A
Operating Junction Temperature Range	T_J	−40 to +125	°C
Storage Temperature Range	T_{stg}	−40 to +150	°C

MAXIMUM RATINGS † (T_J = 25°C unless otherwise noted)

Rating	Symbol	Max	Unit
Thermal Resistance, Junction−to−Case	$R_{\theta JC}$	2.0	°C/W
Maximum Lead Temperature for Soldering Purposes 1/8″ from Case for 10 Seconds	T_L	260	°C

†Indicates JEDEC Registered Data

Stresses exceeding Maximum Ratings may damage the device. Maximum Ratings are stress ratings only. Functional operation above the Recommended Operating Conditions is not implied. Extended exposure to stresses above the Recommended Operating Conditions may affect device reliability.

1. V_{DRM} and V_{RRM} for all types can be applied on a continuous basis. Ratings apply for zero or negative gate voltage; however, positive gate voltage shall not be applied concurrent with negative potential on the anode. Blocking voltages shall not be tested with a constant current source such that the voltage ratings of the devices are exceeded.

ON Semiconductor®

http://onsemi.com

SCRs
12 AMPERES RMS
50 thru 800 VOLTS

MARKING DIAGRAM

TO−220AB
CASE 221A
STYLE 3

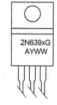

2N639xG
AYWW

2N639x = Device Code
           x = 4, 5, 7, or 9
G   = Pb−Free Package
A   = Assembly Location
Y   = Year
WW   = Work Week

PIN ASSIGNMENT	
1	Cathode
2	Anode
3	Gate
4	Anode

ORDERING INFORMATION

See detailed ordering and shipping information in the package dimensions section on page 4 of this data sheet.

♥ Semiconductor Components Industries, LLC, 2012
September, 2012 − Rev. 9

1

Publication Order Number:
2N6394/D

ELECTRICAL CHARACTERISTICS (T_C = 25°C unless otherwise noted.)

Characteristic		Symbol	Min	Typ	Max	Unit
OFF CHARACTERISTICS						
†Peak Repetitive Forward or Reverse Blocking Current (V_{AK} = Rated V_{DRM} or V_{RRM}, Gate Open)	T_J = 25°C	I_{DRM}, I_{RRM}	–	–	10	μA
	T_J = 125°C		–	–	2.0	mA
ON CHARACTERISTICS						
†Peak Forward On-State Voltage (Note 2) (I_{TM} = 24 A Peak)		V_{TM}	–	1.7	2.2	V
†Gate Trigger Current (Continuous dc) (V_D = 12 Vdc, R_L = 100 Ohms)		I_{GT}	–	5.0	30	mA
†Gate Trigger Voltage (Continuous dc) (V_D = 12 Vdc, R_L = 100 Ohms)		V_{GT}	–	0.7	1.5	V
Gate Non-Trigger Voltage (V_D = 12 Vdc, R_L = 100 Ohms, T_J = 125°C)		V_{GD}	0.2	–	–	V
†Holding Current (V_D = 12 Vdc, Initiating Current = 200 mA, Gate Open)		I_H	–	6.0	50	mA
Turn-On Time (I_{TM} = 12 A, I_{GT} = 40 mAdc, V_D = Rated V_{DRM})		t_{gt}	–	1.0	2.0	μs
Turn-Off Time (V_D = Rated V_{DRM}) (I_{TM} = 12 A, I_R = 12 A)		t_q	–	15	–	μs
(I_{TM} = 12 A, I_R = 12 A, T_J = 125°C)			–	35	–	
DYNAMIC CHARACTERISTICS						
Critical Rate-of-Rise of Off-State Voltage Exponential (V_D = Rated V_{DRM}, T_J = 125°C)		dv/dt	–	50	–	V/μs

†Indicates JEDEC Registered Data
1. Pulse Test: Pulse Width ≤ 300 μsec, Duty Cycle ≤ 2%.

Voltage Current Characteristic of SCR

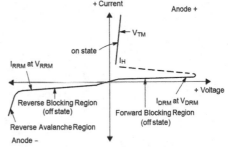

Symbol	Parameter
V_{DRM}	Peak Repetitive Off State Forward Voltage
I_{DRM}	Peak Forward Blocking Current
V_{RRM}	Peak Repetitive Off State Reverse Voltage
I_{RRM}	Peak Reverse Blocking Current
V_{TM}	Peak On State Voltage
I_H	Holding Current

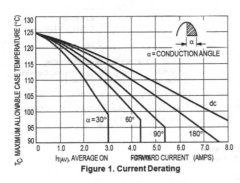

Figure 1. Current Derating

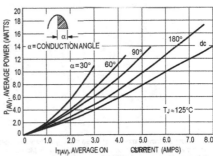

Figure 2. Maximum On-State Power Dissipation

2N6071A/B Series

Preferred Device

Sensitive Gate Triacs

Silicon Bidirectional Thyristors

Designed primarily for full —wave control applications, such as light dimmers, motor controls, heating controls and power supplies; or wherever full —wave silicon gate controlled solid—state devices needed. Triac type thyristors switch from a blocking to a conducting state for either polarity of applied anode voltage with positive or negative gate triggering.

Features

- Sensitive Gate Triggering Uniquely Compatible for Direct Coupling to TTL, HTL, CMOS and Operational Amplifier Integrated Circuit Logic Functions
- Gate Triggering: 4 Mode - 2N6071A, B; 2N6073A, B; 2N6075A, B
- Blocking Voltages to 600 V
- All Diffused and Glass Passivated Junctions for Greater Parameter Uniformity and Stability
- Small, Rugged, Thermopad Construction for Low Thermal Resistance, High Heat Dissipation and Durability
- Device Marking: Device Type, e.g., 2N6071A, Date Code

ON Semiconductor®

http://onsemi.com

TRIACS
4.0 A RMS, 200 - 600 V

MT2 o————————o MT1
o G

REAR VIEW SHOW TAB

TO-225
CASE 077
STYLE 5

3
2 1

MARKING DIAGRAM

1. Cathode
2. Anode
3. Gate

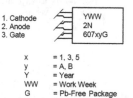

	YWW
	2N
	607xyG

x	= 1, 3, 5
y	= A, B
Y	= Year
WW	= Work Week
G	= Pb-Free Package

ORDERING INFORMATION

See detailed ordering and shipping information in the package dimensions section on page 7 of this data sheet.

Preferred devices are recommended choices for future use and best overall value.

*For additional information on our Pb-Free strategy and soldering details, please download the ON Semiconductor Soldering and Mounting Techniques Reference Manual, SOLDERRM/D.

♥ Semiconductor Components Industries, LLC, 2008
March, 2008 - Rev. 8

1

Publication Order Number:
2N6071/D

MAXIMUM RATINGS (T_J = 25°C unless otherwise noted)

Rating	Symbol	Value	Unit

*Peak Repetitive Off Voltage (Note 1)
(T_J = —40 to 110°C, Sine Wave, 50 to 60 Hz, Gate Open)
    2N6071A,B
    2N6073A,B
    2N6075A,B
	V_{DRM}, V_{RRM}	
		
		200
		400
		600
			V

*On RMS Current (T_C = 85°C) Full Cycle Sine Wave 50 to 60 Hz	$I_{T(RMS)}$	4.0	A
*Peak Non-repetitive Surge Current (One Full cycle, 60 Hz, T_J = +110°C)	I_{TSM}	30	A
Circuit Fusing Considerations (t = 8.3 ms)	I^2t	3.7	A^2s
*Peak Gate Power (Pulse Width ≤ 1.0 µs, T_C = 85°C)	P_{GM}	10	W
*Average Gate Power (t = 8.3 ms, T_C = 85°C)	$P_{G(AV)}$	0.5	W
*Peak Gate Voltage (Pulse Width ≤ 1.0 µs, T_C = 85°C)	V_{GM}	5.0	V
*Operating Junction Temperature Range	T_J	-40 to +110	°C
*Storage Temperature Range	T_{stg}	-40 to +150	°C
Mounting Torque (6 Sx2 Screw) (Note 2)	–	8.0	in. lb.

Stresses exceeding Maximum Ratings may damage the device. Maximum Ratings are stress ratings only. Functional operation above the Recommended Operating Conditions is not implied. Extended exposure to stresses above the Recommended Operating Conditions may affect device reliability.

1. V_{DRM} and V_{RRM} for all types can be applied on a continuous basis. Blocking voltages shall not be tested with a constant current source such that the voltage ratings of the devices are exceeded.
2. Torque rating applies with use of a compression washer. Mounting torque in excess of 6 in. lb. does not appreciably lower case thermal resistance. Main terminal 2 and heatsink contact pad are common.

THERMAL CHARACTERISTICS

Characteristic	Symbol	Max	Unit
*Thermal Resistance, Junction-to-Case	$R_{\theta JC}$	3.5	°C/W
Thermal Resistance, Junction-to-Ambient	$R_{\theta JA}$	75	°C/W
Maximum Lead Temperature for Soldering Purposes 1/8″ from Case for 10 Seconds	T_L	260	°C

*Indicates JEDEC Registered Data.

2N3055AG (NPN),
MJ15015G (NPN),
MJ15016G (PNP)

Complementary Silicon High-Power Transistors

These PowerBase complementary transistors are designed for high power audio, stepping motor and other linear applications. These devices can also be used in power switching circuits such as relay or solenoid drivers, dc–to–dc converters, inverters, or for inductive loads requiring higher safe operating area than the 2N3055.

Features
- High Current–Gain – Bandwidth
- Safe Operating Area
- These Devices are Pb–Free and are RoHS Compliant*

ON Semiconductor®

http://onsemi.com

**15 AMPERE
COMPLEMENTARY SILICON
POWER TRANSISTORS
60, 120 VOLTS – 115, 180 WATTS**

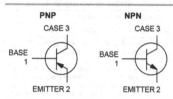

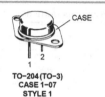

TO–204 (TO–3)
CASE 1–07
STYLE 1

MAXIMUM RATINGS (Note 1)

Rating	Symbol	Value	Unit
Collector–Emitter Voltage 2N3055AG MJ15015G, MJ15016G	V_{CEO}	60 120	Vdc
Collector–Base Voltage 2N3055AG MJ15015G, MJ15016G	V_{CBO}	100 200	Vdc
Collector–Emitter Voltage Base Reversed Biased 2N3055AG MJ15015G, MJ15016G	V_{CEV}	100 200	Vdc
Emitter–Base Voltage	V_{EBO}	7.0	Vdc
Collector Current – Continuous	I_C	15	Adc
Base Current	I_B	7.0	Adc
Total Device Dissipation @ T_C = 25 C 2N3055AG MJ15015G, MJ15016G	P_D	115 180	W W
Derate above 25 C 2N3055AG MJ15015G, MJ15016G		0.65 1.03	W/°C W/°C
Operating and Storage Junction Temperature Range	T_J, T_{stg}	−65 to +200	C

Stresses exceeding Maximum Ratings may damage the device. Maximum Ratings are stress ratings only. Functional operation above the Recommended Operating Conditions is not implied. Extended exposure to stresses above the Recommended Operating Conditions may affect device reliability.
1. Indicates JEDEC Registered Data. (2N3055A)

THERMAL CHARACTERISTICS

Characteristics	Symbol	Max	Max	Unit
Thermal Resistance, Junction–to–Case	$R_{\theta JC}$	1.52	0.98	C/W

MARKING DIAGRAMS

2N3055AG
AYWW
MEX

MJ1501xG
AYWW
MEX

2N3055A = Device Code
MJ1501x = Device Code
    x = 5 or 6
G = Pb–Free Package
A = Assembly Location
Y = Year
WW = Work Week MEX
Origin = Country of

ORDERING INFORMATION

See detailed ordering and shipping information in the package dimensions section on page 5 of this data sheet.

*For additional information on our Pb–Free strategy and soldering details, please download the ON Semiconductor Soldering and Mounting Techniques Reference Manual, SOLDERRM/D.

♥ Semiconductor Components Industries, LLC, 2013
September, 2013 – Rev.7

Publication Order Number:
2N3055A/D

TIP31G, TIP31AG, TIP31BG, TIP31CG (NPN), TIP32G, TIP32AG, TIP32BG, TIP32CG (PNP)

Complementary Silicon Plastic Power Transistors

Designed for use in general purpose amplifier and switching applications.

Features

- High Current Gain – Bandwidth Product
- Compact TO–220 Package
- These Devices are Pb–Free and are RoHS Compliant*

MAXIMUM RATINGS

Rating	Symbol	Value	Unit
Collector – Emitter Voltage TIP31G, TIP32G TIP31AG, TIP32AG TIP31BG, TIP32BG TIP31CG, TIP32CG	V_{CEO}	40 60 80 100	Vdc
Collector–Base Voltage TIP31G, TIP32G TIP31AG, TIP32AG TIP31BG, TIP32BG TIP31CG, TIP32CG	V_{CB}	40 60 80 100	Vdc
Emitter–Base Voltage	V_{EB}	5.0	Vdc
Collector Current – Continuous	I_C	3.0	Adc
Collector Current – Peak	I_{CM}	5.0	Adc
Base Current	I_B	1.0	Adc
Total Power Dissipation @ T_C = 25°C Derate above 25°C	P_D	40 0.32	W W/°C
Total Power Dissipation @ T_A = 25°C Derate above 25°C	P_D	2.0 0.016	W W/°C
Unclamped Inductive Load Energy (Note 1)	E	32	mJ
Operating and Storage Junction Temperature Range	T_J, T_{stg}	−65 to +150	°C

Stresses exceeding those listed in the Maximum Ratings table may damage the device. If any of these limits are exceeded, device functionality should not be assumed, damage may occur and reliability may be affected.

1. I_C = 1.8 A, L = 20 mH, P.R.F. = 10 Hz, V_{CC} = 10 V, R_{BE} = 100 fi

THERMAL CHARACTERISTICS

Characteristic	Symbol	Max	Unit
Thermal Resistance, Junction–to–Ambient	$R_{\theta JA}$	62.5	°C/W
Thermal Resistance, Junction–to–Case	$R_{\theta JC}$	3.125	°C/W

*For additional information on our Pb–Free strategy and soldering details, please download the ON Semiconductor Soldering and Mounting Techniques Reference Manual, SOLDERRM/D.

♥ Semiconductor Components Industries, LLC, 2015
September, 2015 – Rev.16

Publication Order Number:
TIP31A/D

ON Semiconductor®

www.onsemi.com

3 AMPERE POWER TRANSISTORS COMPLEMENTARY SILICON 40–60–80–100 VOLTS, 40 WATTS

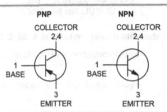

PNP	NPN
COLLECTOR 2,4	COLLECTOR 2,4
1 BASE	1 BASE
3 EMITTER	3 EMITTER

TO–220
CASE 221A
STYLE 1

MARKING DIAGRAM

TIP3xxG
AYWW

TIP3xx	= Device Code
xx	= 1, 1A, 1B, 1C, 2, 2A, 2B, 2C,
A	= Assembly Location
Y	= Year
WW	= Work Week
G	Pb–Free Package

ORDERING INFORMATION

See detailed ordering and shipping information on page 6 of this data sheet.

COMPATIBLE WITH STANDARD TTL INTEGRATED CIRCUITS

Gallium-Arsenide-Diode Infrared Source Optically Coupled to a Silicon npn Phototransistor

High Direct-Current Transfer Ratio

High-Voltage Electrical Isolation
1.5-kV, 2.5-kV, or 3.55-kV Rating

High-Speed Switching

$t_r = 7$ μs, $t_f = 7$ μs Typical

Typical Applications Include Remote Terminal Isolation, SCR and Triac Triggers, Mechanical Relays and Pulse Transformers

Safety Regulatory Approval
UL/CUL, File No. E65085

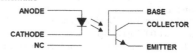

DCJ† OR 6-TERMINAL DUAL-IN-LINE PACKAGE
(TOP VIEW)

ANODE	1	6	BASE
CATHODE	2	5	COLLECTOR
NC	3	4	EMITTER

†4N35 only
NC – No internal connection

schematic

ANODE ——————— BASE
CATHODE ——————— COLLECTOR
NC ——————— EMITTER

absolute maximum ratings at 25°C free-air temperature (unless otherwise noted)†

Input-to-output peak voltage (8-ms half sine wave): 4N35	3.55 kV
4N36	2.5 kV
4N37	1.5 kV
Input-to-output root-mean-square voltage (8-ms half sine wave): 4N35	2.5 kV
4N36	1.75 kV
4N37	1.05 kV
Collector-base voltage	70 V
Collector-emitter voltage (see Note 1)	30 V
Emitter-base voltage	7 V
Input-diode reverse voltage	6 V
Input-diode forward current: Continuous	60 mA
Peak (1 μs, 300 pps)	3 A
Phototransistor continuous collector current	100 mA
Continuous total power dissipation at (or below) 25°C free-air temperature:	
Infrared-emitting diode (see Note 2)	100 mW
Phototransistor (see Note 3)	300 mW
Continuous power dissipation at (or below) 25°C lead temperature:	
Infrared-emitting diode (see Note 4)	100 mW
Phototransistor (see Note 5)	500 mW
Operating temperature range, T_A	−55°C to 100°C
Storage temperature range, T_{stg}	−55°C to 150°C
Lead temperature 1,6 mm (1/16 inch) from case for 10 seconds	260°C

† Stresses beyond those listed under "absolute maximum ratings" may cause permanent damage to the device. These are stress ratings only, and functional operation of the device at these conditions is not implied. Exposure to absolute-maximum-rated conditions for extended periods may affect device reliability.

NOTES:
1. This value applies when the base-emitter diode is open-circulated.
2. Derate linearly to 100°C free-air temperature at the rate of 1.33 mW/°C.
3. Derate linearly to 100°C free-air temperature at the rate of 4 mW/°C.
4. Derate linearly to 100°C lead temperature at the rate of 1.33 mW/°C. Lead temperature is measured on the collector lead 0.8 mm (1/32 inch) from the case.
5. Derate linearly to 100°C lead temperature at the rate of 6.7 mW/°C.

Please be aware that an important notice concerning availability, standard warranty, and use in critical applications of Texas Instruments semiconductor products and disclaimers thereto appears at the end of this data sheet.

TEXAS INSTRUMENTS

POST OFFICE BOX 655303 □ DALLAS, TEXAS

1

Powerex, Inc., 173 Pavilion Ln, Youngwood, PA 15697-1800 (724)925-7272 WWW.PWRX.COM

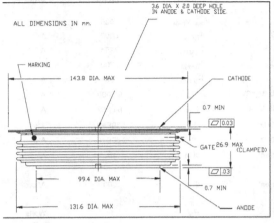

ALL DIMENSIONS IN mm.

3.6 DIA. X 2.0 DEEP HOLE IN ANODE & CATHODE SIDE.

MARKING

143.8 DIA. MAX

CATHODE

0.7 MIN

0.03

GATE 26.9 MAX (CLAMPED)

.03

99.4 DIA. MAX

0.7 MIN

131.6 DIA. MAX

ANODE

The TDS5 is a low voltage, high current, thin pack disc SCR employing a amplifing gate structure. This thin package provides greater cooling thus maximizing high current performance. The amplifing gate design allows the SCR to be reliably operated at high di/dt and dv/dt conditions in various phase control applications.

FEATURES:
Low On-State Voltage
High di/dt Capability
High dv/dt Capability
Hermetic Ceramic Package
Excellent Surge and I^2t Ratings

APPLICATIONS:
DC Power Supplies
Motor Controls
Plating Rectifiers

ORDERING INFORMATION

Select the complete 12 digit Part Number using the table below.
EXAMPLE: TDS520503DH is a 2000V-5000A SCR with 200ma IGT and 12 inch gate and cathode potential leads.

PART	Voltage Rating	Voltage Code	Current Rating	Current Code	Turn-Off	Gate	Leads
	V_{DRM}-V_{RRM}		I_{tavg}		T_q	I_{GT}	
TDS5	2000	**20**	5000	**50**	**0**	**3**	
	1800	**18**					
	1600	**16**			500us	200ma	12"
	1400	**14**			(typ.)	(max)	
	1200	**12**					

Revised: 4/23/2009

Page 1

Absolute Maximum Ratings

Characteristic	Symbol	Rating	Units
Repetitive Peak Voltage	V_{DRM}-V_{RRM}	2000	Volts
Average On-State Current, T_C=68°C	$I_{T(Avg.)}$	5000	A
RMS On-State Current, T_C=68°C	$I_{T(RMS)}$	7854	A
Average On-State Current, T_C=55°C	$I_{T(Avg.)}$	5700	A
RMS On-State Current, T_C=55°C	$I_{T(RMS)}$	8954	A
Peak One Cycle Surge Current, 60Hz, V_R=0V	I_{TSM}	90,000	A
Peak One Cycle Surge Current, 50Hz, V_R=0V	I_{TSM}	84,852	A
Fuse Coordination I^2t, 60Hz	I^2t	3.38E+07	A^2s
Fuse Coordination I^2t, 50Hz	I^2t	3.60E+07	A^2s
Critical Rate-of-Rise of On-State Current Repetitive	di/dt	100	A/us
Critical Rate-of-Rise of On-State Current Non-Repetitive	di/dt	300	A/us
Peak Gate Power, 100us	P_{GM}	16	Watts
Average Gate Power	$P_{G(avg)}$	5	Watts
Operating Temperature	Tj	-40 to+125	°C
Storage Temperature	$T_{Stg.}$	-50 to+150	°C
Approximate Weight		6.5	lb
		2.95	Kg
Mounting Force		16,000-20,000	lbs
		71.2 - 89.0	KNewtons

Information presented is based upon manufacturers testing and projected
capabilities. This information is subject to change without notice. The
manufacturer makes no claim as to suitability for use, reliability, capability or
future availability of this product.

Powerex, Inc., 173 Pavilion Lane, Youngwood, Pennsylvania 15697 (724) 925-7272
www.pwrx.com

PM50B4L1C060

Photo Voltaic IPM
H-Bridge
50 Amperes/600 Volts

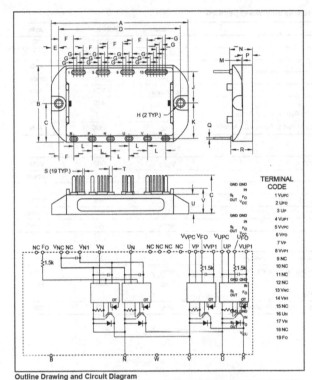

Outline Drawing and Circuit Diagram

Dim.	Inches	Millimeters	Dim.	Inches	Millimeters
A	3.54	90.0	L	0.47	12.0
B	1.97	50.0	M	0.012	0.3
C	0.98	25.0	N	0.57	14.6
D	3.15	80.0	P	0.26	6.7
E	0.20	5.0	Q	0.02	0.5
F	0.39	10.0	R	0.56	14.2
G	0.08	2.0	S	0.02 Sq.	0.5 Sq.
H	0.17 Dia.	4.3 Dia.	T	0.08	2.0
J	0.81	20.5	U	0.51	13.0
K	0.91	23.0	V	0.65	16.5

TERMINAL CODE

1	VUPC
2	UFO
3	UP
4	VUP1
5	VVPC
6	VFO
7	VP
8	VVP1
9	NC
10	NC
11	NC
12	NC
13	VNC
14	VN1
15	NC
16	UN
17	VN
18	NC
19	FO

Description:

Powerex Intellimod™ Photo Voltaic Intelligent Power Modules are isolated base modules designed for single phase power switching applications. Built-in control circuits provide optimum gate drive and protection for the IGBT and free-wheel diode power devices.

Features:

☐ Complete Output Power Circuit
☐ Gate Drive Circuit
☐ Protection Logic
  – Short Circuit
  – Over Temperature Using On-chip Temperature Sensing
  – Under Voltage
☐ Low Loss Using Full Gate CSTBT IGBT Chip

Applications:

☐ PV Inverters
☐ PV UPS
☐ PV Power Supplies

Ordering Information:

Example: Select the complete part number from the table below -i.e. PM50B4L1C060 is a 600V, 50 Ampere PV-IPM.

Type	Current Rating Amperes	VCES Volts (x10)
PM	50	60

03/11 Rev. 0

1

Powerex, Inc., 173 Pavilion Lane, Youngwood, Pennsylvania 15697 (724) 925-7272 www.pwrx.com

PM50B4L1C060
Photo Voltaic IPM
H-Bridge
50 Amperes/600 Volts

Absolute Maximum Ratings, T_j = 25°C unless otherwise specified

Characteristics	Symbol	PM50B4L1C060	Units
Power Device Junction Temperature	T_j	-20 to 150	°C
Storage Temperature	T_{stg}	-40 to 125	°C
Mounting Torque, M4 Mounting Screws (Typical)	—	15	in-lb
Module Weight (Typical)	—	135	Grams
Supply Voltage, Surge (Applied between P-N)	$V_{CC(surge)}$	500	Volts
Operation of Short Circuit Protections	$V_{CC(prot.)}$	450	Volts
(Applied between P-N, V_D = 13.5 ~ 16.5V, Inverter Part, T_j = 125°C Start)			
Isolation Voltage (60Hz, Sinusoidal, RMS, Charged Part to Base, AC 1 Minute)	V_{ISO}	2500	Volts

Inverter Part

Collector-Emitter Voltage (V_D = 15V, V_{CIN} = 15V)	V_{CES}	600	Volts
Collector Current (T_C = 25°C)	I_C	50	Amperes
Collector Current (Pulse)	I_{CRM}	100	Amperes
Total Power Dissipation (T_C = 25°C)	P_{tot}	168	Watts
Emitter Current (T_C = 25°C, FWDi Current)	I_E	50	Amperes
Emitter Current (Pulse, FWDi Current)	I_{ERM}	100	Amperes

Control Part

Supply Voltage (Applied between V_{UP1}-V_{UPC}, V_{VP1}-V_{VPC}, V_{N1}-V_{NC})	V_D	20	Volts
Input Voltage (Applied between U_P-V_{UPC}, V_P-V_{VPC}, U_N-V_N-W_N-Br-V_{NC})	V_{CIN}	20	Volts
Fault Output Supply Voltage	V_{FO}	20	Volts
(Applied between U_{FO}-V_{UPC}, V_{FO}-V_{VPC}, F_O-V_{NC})			
Fault Output Supply Current (Sink Current at U_{FO}, V_{FO}, F_O Terminals)	I_{FO}	20	mA

2

03/11 Rev. 0

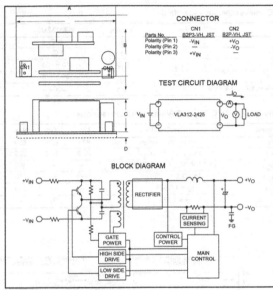

CONNECTOR

Parts No.	CN1 B2P3-VH, JST	CN2 B2P-VH, JST
Polarity (Pin 1)	-V$_{IN}$	+V$_O$
Polarity (Pin 2)	—	-V$_O$
Polarity (Pin 3)	+V$_{IN}$	—

TEST CIRCUIT DIAGRAM

BLOCK DIAGRAM

Outline Drawing and Circuit Diagram

Dimensions	Inches	Millimeters
A	3.54	90.0
B	2.16	55.0
C	1.38	35.0
D	0.20	5.0

Description:
VLA312-2425 is an isolated DC-DC converter designed for industrial equipment. It is designed to convert a rectified line voltage ranging from 475 to 850V DC into 24V DC. Total output power is 25W.

Features:
- Input Voltage Range: 475V to 850V DC
- Output: +24V, 1.05A (Output Power: 25.2W)
- Electrical Isolation Voltage Between Input and Output: 2500 V$_{rms}$ for 1 Minute
- Over-current Protection (Auto Resumption)
- Over-voltage Protection

Application:
On-board pre-regulator for industrial control equipment.

Powerex, Inc., 173 Pavilion Lane, Youngwood, Pennsylvania 15697 (724) 925-7272

VLA312-2425
Isolated DC-DC Converter

Absolute Maximum Ratings, T_a = 25°C unless otherwise specified

Characteristics	Symbol	VLA312-2425	Units
Input Voltage	V_{IN}	850	Volts
Output Current	I_O	1.05	A
Operating Temperature (No Condensation)*1	T_{opr}	-10 ~ +55	°C
Storage Temperature (No Condensation)	T_{stg}	-20 ~ +75	°C
Input-Output Isolation Voltage (Sine Wave Voltage, 60Hz, 1 Minute)	V_{ISO}	2500	V_{rms}

Electrical and Mechanical Characteristics, T_a = 25°C, V_{IN} = 680V unless otherwise noted

Characteristics	Symbol	Test Conditions	Min.	Typ.	Max.	Units
Input Voltage	V_{IN}	Recommended Range	475	680	850	Volts
Output Voltage 1	V_O	I_O = 0.1 ~ 1.05A	22.8	24.0	25.2	Volts
Input Regulation	R_{eg-I}	I_O = 1.05A, V_{IN} = 475 ~ 850V	—	—	100	mV
Load Regulation	R_{eg-L}	I_O = 0.1 ~ 1.05A	—	—	150	mV
Ripple Voltage	V_{P-P}	I_O = 1.05A*2	—	—	240	mV
Efficiency	η	I_O = 1.05A	—	77	—	%

*1 Please refer to de-rating characteristics.
*2 Not including spike noise.

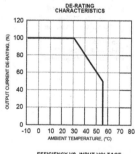

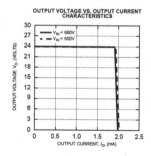

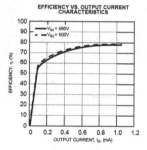

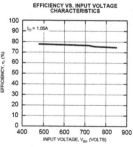

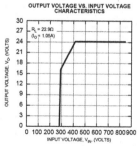

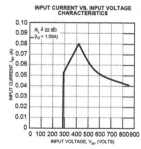

2

05/09